Looking Back

NIAGARA'S GORGE BRIDGES

MARVELS OF ENGINEERING

Dedicated to the Niagara Falls Bridge Commission, which, since 1938, has operated as a public authority maintaining three great bridges over the Niagara Gorge, linking Canada and the United States in a friendly and cooperative enterprise for the benefit of all.

Looking Back

Niagara's Gorge Bridges

Marvels of Engineering

Paul E. Lewis

Looking Back Press

Published by Looking Back Press
An Imprint of Vanwell Publishing Limited
1 Northrup Crescent, P.O. Box 2131
St. Catharines, ON L2R 7S2
For all general information contact Looking Back Press at:
Telephone 905-937-3100 ext. 829
Fax 905-937-1760
E-Mail vanessa.mclean@vanwell.com

For customer service and orders:
Toll-free 1-800-661-6136

Printed in Canada

Library and Archives Canada Cataloguing in Publication

Lewis, Paul E. (Paul Eugene), 1940-
Niagara's Gorge bridges : marvels of engineering / Paul E. Lewis.

ISBN 978-1-55068-925-9

1. Bridges—Niagara Gorge (N.Y. and Ont.)—History. 2. Bridges—Niagara Gorge (N.Y. and Ont.)—Design and construction—History. 3. Niagara Gorge (N.Y. and Ont.)—History. I. Title.

FC3095.N5Z533 2008 624.2'0971338 C2008-903458-9

Cover Photo: Construction activity at the building site of the Niagara Railway Steel Arch Bridge in 1896. (NFBC Archives)

Contents

Acknowledgments 6

Introduction 7

1. Niagara Suspension Bridge 1848–1855 9

2. Railway Suspension Bridge 1855–1897 17

3. Niagara Railway Steel Arch Bridge 1897 29

4. Niagara Cantilever Bridge 1883–1925 45

5. Michigan Central Railway Arch Bridge 1925 51

6. Upper Suspension Bridge 1869–1899 57

7. Upper Steel Arch Bridge 1898–1938 67

8. Rainbow Bridge 1941 89

9. Lewiston-Queenston Suspension Bridge 1851–1864 109

10. Lewiston-Queenston Suspension Bridge 1899–1962 113

11. Lewiston-Queenston Steel Arch Bridge 1962 121

12. Goat Island Bridges 143

Acknowledgements

This project began with the uncovering of a marvelous cache of engineers' photographs from the building of the Steel Arch bridges in the late 1890s. These were all stored in the old offices area of the Niagara Falls Bridge Commission. This led to a project at the Bridge Commission to archive their collections in a more systematic fashion. I am most grateful to the Commission for the opportunity to do this. I am also most grateful for the encouragement and suggestions along the way offered by the Commission's General Manager Mr. Tom Garlock, and by the helpfulness of his able executive secretary Sandra Phillips. It seemed fitting that the 70th anniversary of the establishment of the Bridge Commission could be celebrated with this publication.

I would also acknowledge a great debt to the late George Siebel, author of numerous publications on the Niagara area, including his 1990 *Bridges Over The Niagara Gorge*. His research was excellent, and that particular work made an outstanding contribution to bridge history.

Myriad other photographs were required, of course, and my sincere thanks are extended to a number of institutions and individuals who made their collections freely available for research and publication. They are: Andrew Porteus, Manager of Adult Reference and Information Services, Niagara Falls Public Library; Maureen Fennie, Assistant Director and Manager of the Local History Department at the Niagara Falls, New York, Public Library; the Special Collections at the St. Catharines Public Library; George Bailey; Sherman Zavitz; and especially John Burtniak, whose postcard collection is so remarkable. Every effort has been made to acknowledge the source of each of the photos so that future researchers can find them again.

Special thanks must also be extended to H. Jane Lewis of the small but profoundly professional editing firm Text Etc. of Guelph. Her time and suggestions were invaluable.

And to my wife Janet, for her patience and remarkable computer expertise as this book came together, must go my everlasting gratitude.

I am grateful also to Ben Kooter, and Vanessa McLean, Managing Editor of Looking Back Press. Their support for, and interest in, local history is an inspiration to publishers everywhere.

Introduction

It is the engineers of this world who have dreamed, designed, and shaped our built environment. To conceive and design any product, be it a safety pin, a computer, a space shuttle, or a bridge, requires flights of imagination and confident responses to human needs. Engineers' imaginations are then coupled with originality, broad scientific knowledge, and mathematical skill. All the parts must be known and all must fit together with exact precision. Since the end products are normally constructed by teams, engineers must also then combine their skills with judgment, understanding, and indeed courage. Examining any engineering achievement that blends science, art, mathematics, and function should elicit a thrill in those of us who are not engineers. While we often simply pay our money and take the end product or structure for granted, we should pause now and again to reflect on the efforts required to bring us the abundance of products and structures we enjoy. The efforts of the engineers are never static. Each engineered item is affected by time and changing needs, and each one influences the next. Their failures, too, in the end bring us both better engineers and better products. No better example of engineers at their finest in their particular eras can be found than in an examination of their work in the construction of Niagara's gorge bridges.

There is something both simple and profound about all bridges. On one hand, they cross bodies of water both large and small, and get us on our way. On the other hand, they can be seen as metaphors for human connections as they represent invitations to peaceful access to the other side. They begin with dreams, and are opened with celebrations. While all bridges are admired by the generation that sees their completion, none are permanent. All are affected by pollution, neglect, unintended abuse, wear and tear and old age, and all eventually give way to their successors, having left a legacy to the environment and to their users. Bridges capture the imaginations of painters, poets, architects, writers, and photographers. They are technological and aesthetic statements proving that society is better served by their existence. Each bridge offers new challenges and each reflects the changing needs of communication, theories of design, and advances in technology.

The bridges over the Niagara Gorge are indeed marvels of the engineering profession both past and present. The bridges closest to the Falls have always provided a balcony to the natural wonder of the cataract. Those at the Whirlpool have always favoured cross-border convenience. The bridges at Lewiston-Queenston have served more as highways of commerce. The Goat Island structures benefit the tourists. Historically, the construction of each bridge tells a story about the flowering of engineering as a profession from the mid-19th century to the mid-20th. All were built by masterminds with large supporting casts. Those which stand today are certainly worthy of more than just a second look.

Paul E. Lewis
St. Catharines, Ontario
May 2008

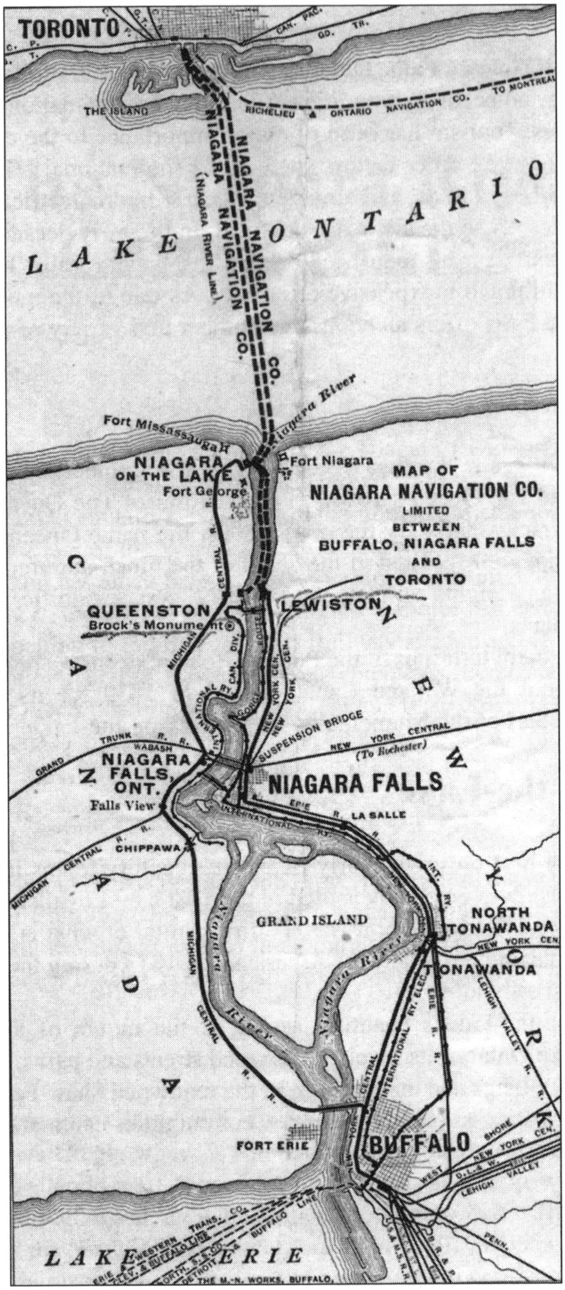

Guide to Abbreviations used to identify photographic sources

NFBC Archives	Niagara Falls Bridge Commission Archives
NF PL	Niagara Falls Public Library
NF NY PL	Niagara Falls, New York, Public Library

One

Niagara Suspension Bridge

The builder who first bridged Niagara's gorge.
Before he swung his cable, shore to shore,
Sent out across the gulf his venturing kite
Bending a slender cord for unseen hands
To grasp upon the further cliff and draw
A greater cord, and then a greater yet;
Till at last across the chasm swung
The cable – then the mighty bridge in air!

William Hamilton Merritt (1793–1862) was the instigator of the first bridge over the Niagara Gorge. His portrait above appears in numerous texts, including the biography completed by his son in 1875.

The poem above by American poet Edwin Markham (1852–1940) was a tribute to thirteen-year-old Homan Walsh, who flew his kite across the Niagara Gorge in January 1848. It carried the first string connection between Canada and the United States, and marked the beginning of construction of the first bridge to link the two countries.

Charles Ellet (1810–1862) Bridge Builder

Charles Ellet was born in 1810 at Penn's Manor, Pennsylvania, the son of a Quaker farmer. Very early in his life he displayed an unusual mathematical ability, a phenomenal memory and a gift for languages. He left home at seventeen and began his career as a surveyor and assistant engineer on the Chesapeake & Ohio Canal. During that time he taught himself French and saved enough money to go to Paris to study engineering at the École Polytechnique. This sojourn also allowed him to study other engineering works taking place at the time in Germany, France, and England. Ellet returned home in 1832, the first American with a European education in engineering.

Various engineering projects followed, primarily on canals, but his main interest came to lie with the design of suspension bridges. His first opportunity to build came in 1841–42 when he constructed America's first long-span wire-cable suspension bridge across the Schuylkill River near Philadelphia. Other bridge designs followed, and by 1845 Ellet was the best-known bridge builder in North America. His reputation and flamboyant style earned him the contract in November 1847 to design and build the first suspension bridge across the Niagara River. He completed his carriage and pedestrian bridge in just over eight months. This success was followed in 1848–49 with his suspension bridge over the Ohio River at Wheeling, West Virginia, then the longest single span in the world at 1010 feet (308 m). In the 1850s he worked on surveys for the U.S. War Department and as chief engineer of the Virginia Central Railroad. Following the outbreak of the Civil War in 1861, Ellet designed and built steam-powered ships to ram Confederate ships on the Mississippi River. In June 1862, while commanding a warship in the Battle of Memphis, Colonel Ellet was struck by a bullet and died shortly after.

Early documents support the fact that it was William Hamilton Merritt of St. Catharines who first conceived the practicability of a bridge across the Niagara Gorge. Merritt (1793–1862) had spearheaded the building of the first Welland Canal, which opened in 1829. He understood the complexity of engineering projects and was well connected with a variety of entrepreneurs on both sides of the border who would finance the undertaking. In the early 1840s he saw the commercial potential of a bridge and he also foresaw the need for a rail crossing in the near future. While his accomplishments in government (particularly in his forward thinking ideas on reciprocal trade), in banking, and in numerous successful enterprises earned him the grudging respect of his contemporaries as a visionary, Merritt earned the modern sobriquet "Father of Canadian Transportation" for his more practical accomplishments in that field. The first bridge over the Niagara Gorge is a testament to but one of these. This statue of Merritt in St. Catharines shows him gazing north along the first and second Welland Canal routes.

Dual charters to build a bridge were secured in 1846 from the State of New York and from the Canadian government. Two private companies were formed: the Niagara Falls International Bridge Company and the Niagara Falls Suspension Bridge Company. Together they elected W. H. Merritt as president. Merritt, the entrepreneur, banker, and member of the Canadian government, insisted that a provision for a railway deck be added to the charters. The companies then selected Charles Ellet to build this first gorge crossing near the present Whirlpool Bridge. The site chosen was the narrowest point, bluff to bluff, stretching nearly 800 feet (244 m) across the gorge. This etching of the whirlpool area, made in the 1850s, gives some indication of the terrain and the distance across the river that Ellet would have to master. (*The English Cricketer's Trip to Canada and the United States*, 1859)

Ellet's first problem was to establish the link across the gorge. It was a friend of Ellet's, local iron-worker Theodore G. Hullett (later Judge Hullett) of Niagara Falls, NY, who suggested a kite flying contest. Five dollars in prize money attracted a large number of kite flyers, but only one—a Niagara Falls, NY, youth named Homan Walsh—was successful. A legend was born that January 1848. As the poem on page nine suggests, the kite string pulled a heavier cord across the gorge, followed by a rope and finally by a wire cable consisting of thirty-six strands of No. 9 wire. Construction was ready to begin on this first cross-border connection. (NF NY PL)

Ellet's contract of November 1847 stipulated May 1849 as the completion date for a combined road and rail bridge. This was far too short a time in which to accomplish this daring engineering feat. Whether he realized this at the time is unknown, but he would make the most of the contract anyway. With the first cable connection completed, Ellet built two 50-foot (15.2 m) temporary towers on top of the bluffs, one on each side of the gorge. Then a long 1190-foot (363 m) cable consisting of thirty-six strands of No. 10 wire was strung from the towers and anchored to the rock behind. This was completed by March 1948. (NF NY PL)

A way had to be found to get workmen and materials back and forth across the gorge. To accomplish this, Ellet's friend Theodore Hullett fashioned an iron basket and had it suspended from rollers on the cable. On a cold March 13th morning, to demonstrate his faith in both cable and basket, the self-confident Ellet was winched across and back again, exploiting this historic moment. This feat was accomplished in a short fifteen minutes, to the excitement of the crowds gathered on both bluffs. Nobody made a better show of bridge building! Ellet was the first man to cross the gorge by bridge. His original iron basket, shown above, is on display in the collection of The Buffalo and Erie County Historical Society in Buffalo, NY. (NF NY PL)

Ellet quickly discovered that many people wanted to cross the river in this strange contraption. He promptly charged a fee for the privilege and began generating considerable cash. Even W. H. Merritt crossed this way in March 1848, on one of his inspection tours. Soon after, the Board of Directors gave notice to Ellet that he was to stop ferrying passengers and use it solely for construction purposes. Progress was rapid in the spring and early summer. (NF NY PL)

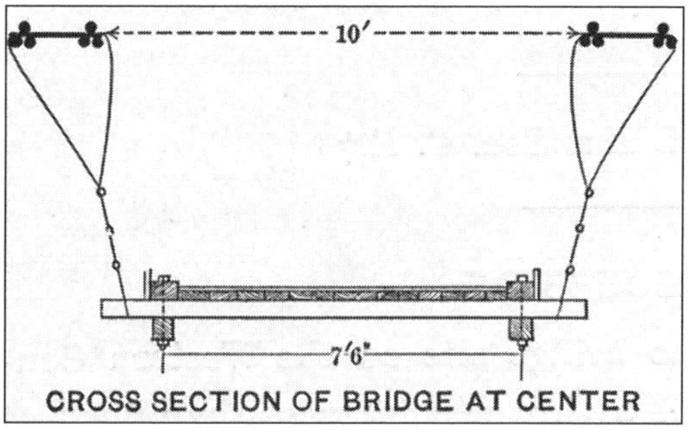

Ellet's basic design is shown in this cross section taken at the center. The four wooden towers were completed in May 1848, and the cables were strung over the tops. Each cable consisted of seventy-two strands of No. 10 wire and was wrapped with resin-coated binding. The bridge had no stiffening trusswork. Instead the 7.5-foot (2.3 m) carriageway hung from suspenders attached to the four main cables, a pair from each tower. Guy wires were attached to the cliffs below. The bridge was 762 feet (232 m) long. (L. L. Buck, *Report on the Construction of The Steel Arch Bridge*, 1899)

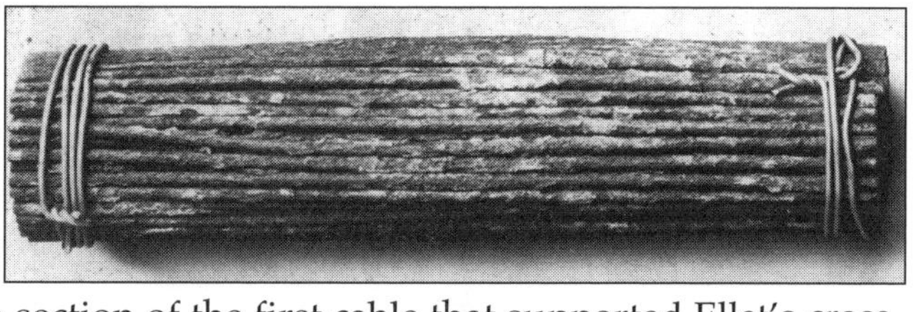

This is a photo claimed to be of a section of the first cable that supported Ellet's crossing in the iron basket on March 13, 1848. It was part of the five hundred tons of wire manufactured by Richard Johnson of Manchester, England. Since the wires are not twisted as in subsequent bridge cables, the claim may well be true. The resin binding that wrapped all of Ellet's cables has been removed. (NFBC Archives)

Ellet completed his work in the summer of 1848. In July, after a catwalk of oak planks had been strung across, and before the guardrails had been set into place, he boldly rode his horse across the gorge. He repeated the trip in a small carriage, reportedly standing up. His bravado on this occasion created another Niagara Falls legend, one that greatly outlasted his bridge. The bridge officially opened to the public on August 1, and it quickly became a financial success, especially for Ellet who continued to collect fees. A disagreement soon arose with the bridge companies over ownership of the funds. After some tense days and bitter litigation in October, Ellet walked away, never to come back. His bridge swung in the wind, but it was safe and served its purpose both as a novelty for the tourist trade and as the first, albeit elementary and temporary, commercial highway across the Niagara River. (NF PL)

In the middle of the 19th century, the idea of being able to cross the Niagara River on a bridge was awe inspiring. Today we take such things for granted. Then, constructions like the Niagara Suspension Bridge were seen as marvels of the progress of the century. In this pre-photographic era, people were informed about such things by newspapers. Some correspondents managed to capture the excitement of the experience.

from the (Toronto) *Globe*, August 2, 1848

...But on getting on the Bridge with but a board laid on two or three wires on either side...you look down the giant chasm to the green and white current moving so majestically beneath...and as your frail footing trembles to an approaching step or sways to the passing breeze, you know that but an iron thread is between you and the [abyss] below.

from the *Buffalo Daily Republic*, August 25, 1848

...not only is the present structure capable of sustaining heavy two horse carriages, but it will bear up on the centre 70 tons weight, hence there is little cause of fear from our weightiest men. It is true that there is here and there one who estimates their weight too much for so slight an edifice, yet when they place themselves above the rushing waters...they become men again.

This first Niagara Suspension Bridge (Ellet's bridge) served the public for less than six years. Ellet's connection with it ceased soon after its completion. In 1851, the directors of the bridge companies set the wheels in motion for the construction of a railway and carriage bridge, their original intent in 1846. This was a much more ambitious project and would be under the direction of civil engineer John A. Roebling. He would use Ellet's suspension bridge as scaffolding until 1854 when it was no longer needed. It was quickly and quietly dismantled. (NF PL)

On Monday, August 2, 1948, a gala Arch of Friendship centennial program was held to mark one hundred years of uninterrupted service across the Niagara Gorge. Since the first span no longer existed, festivities were held at the Whirlpool Bridge. There were a variety of events at both American and Canadian ends and on the bridge itself. These included the launch of a commemorative postage stamp; a meeting of New York's Lieutenant Governor with the Ontario Premier at the halfway point, where they unveiled a plaque; band concerts; an aerial display by RCAF jet planes; a meeting of Miss Niagara Falls, Ontario, with the Queen of Niagara Falls, New York and, of course, a spectacular fireworks display at ten p.m., staged from a flatcar on the bridge.

In the afternoon, the Canadian and American presidents of the Bridge Company presented five silver dollars to boys in 1840s costumes in memory of Homan Walsh, the boy who first flew a kite across the river in 1848. The feat itself was repeated that day using a helicopter carrying a telephone cable. For all the festivities, the bridge itself was closed to traffic. It was the first time in one hundred years that the crossing between countries was impeded. The commemorative postage stamp appears at right. The plaque reads as follows:

AT THIS SITE STOOD THE FIRST NIAGARA SUSPENSION BRIDGE. BUILT IN 1848 THAT BRIDGE AND ITS THREE SUCCESSORS HAVE GIVEN ONE HUNDRED YEARS OF UNINTERRUPTED SERVICE TO THE PEOPLES OF THE UNITED STATES AND CANADA, AND HAVE STOOD AS MONUMENTS TO THE FRIENDSHIP EXISTING BETWEEN THESE TWO COUNTRIES. THEY HAVE SYMBOLIZED TO ALL THE WORLD THAT NATIONS MAY LIVE AT PEACE WITH ONE ANOTHER.

Unveiling the plaque on August 2, 1948, was the Honourable George C. Drew, Premier of Ontario (1943–1948), with the flag of Ontario (above), and the Honorable Joseph R. Hanley, Lieutenant Governor of the State of New York (1943–1950), with the New York flag and the Stars and Stripes (below). The photos were taken at the middle of the Whirlpool Bridge where the plaque was unveiled. (NFBC Archives)

Two

Railway Suspension Bridge
(Great International Railway Suspension Bridge)
(Niagara Bridge)

This 1859 picture of the Great International Railway Suspension Bridge was reproduced in many versions, both black-and-white and colour. It originated from a line engraving by D. L. Glover, copied from a painting by Danish artist Ferdinand Richardt, who in turn painted it from a daguerreotype. Colour versions were created by Hallmark Cards. The picture shows the intricate network of cables suspended from the bridge's towers. With the Falls in the background, an early Maid of the Mist tour boat is also visible near the left bank above the rapids. At the time of the bridge's opening in 1855, one eloquent journalist praised the structure, suggesting that "the hand of man had enhanced the hand of God who had carved the gorge in the first place." (L. L. Buck, *Report on the Construction of The Steel Arch Bridge*, 1899)

John A. Roebling (1806–1869) Bridge Builder

John Roebling was born in Mühlhausen, Germany, in 1806. He was educated there before attending the Royal Polytechnic Institute in Berlin where he studied architecture, engineering, bridge construction, hydraulics, languages, and philosophy. He graduated in 1826 and began his career working for the government as a road builder. Looking further afield for intellectual stimulation, he immigrated to the United States in 1831 and spent several years farming wild land in the Pittsburgh area. Finding this monotonous, he longed for engineering opportunities. By 1837, he was an assistant engineer in various surveying and canal projects in Pennsylvania.

Seeing canal barges being hauled with endless hemp ropes sparked the idea of wire rope as a cheaper and safer alternative. In 1841, Roebling began making twisted seven-strand wire rope, and in 1842 he perfected and patented his process, which ensured uniform tension on all strands. He used his wire rope in a series of successful suspension aqueducts in the 1840s, all of which gained him valuable experience. In 1849, Roebling moved from the Pittsburgh area to Trenton, New Jersey, where he established himself as the first major wire rope manufacturer in America.

In 1850, Roebling's attention was drawn to Niagara Falls where there was an urgent need to replace Ellet's light carriage and foot bridge (completed in 1848) with one that would sustain rail traffic (see Chapter One). He began building this suspension bridge in 1852. The two bridge decks supported by wire cables were officially opened for traffic in March 1855. A series of successful suspension bridges followed in other locations, although the outbreak of the Civil War ended construction for a couple of years. In 1867, Roebling was chosen as chief engineer and designer for the proposed Brooklyn Suspension Bridge in New York. He prepared the plans and was superintending the initial construction when, in July 1867, his foot was crushed in a pier by a ferryboat. This accident brought on tetanus, which led to his death in July 1869. His son, Washington Roebling, completed that epic project in 1883.

All suspension bridges had dubious reputations. In all of America in the early 1850s, there were only two engineers who had any firm belief in them. Charles Ellet was one, John Roebling the other. Roebling's fundamental belief was that the stiffer and heavier the roadway could be constructed, the more stable the bridge would be. He understood the principles of anchor stays, how to spin his cables, the need for stiffening beams and railings as well as cables to counteract the action of the wind. His trusses would be arranged as a combination of triangles. These, he was convinced, were the means of preventing the vibrations arising from heavy loads on the railbed. This early drawing shows the four cables and the box-like design. (*The Falls of Niagara*, ca. 1880)

Roebling's bridge proposal, submitted in August 1851, was accepted in June 1852 by the Niagara Falls International Bridge Company and the Niagara Falls Suspension Bridge Company. He was then ordered to submit his plans and estimates for a two-floor bridge, "both connected by stiff trusses, so as to form a box, suspended to four wire cables of an aggregate strength equivalent to five times the maximum tension caused by the passage of the heaviest train at 5 miles per hour and the weight of the structure." He contracted to provide all engineering services including design and supervision of the construction. This illustration shows an 1860s train approaching the American side. (*Tugby's Illustrated Guide to Niagara Falls*, 1889)

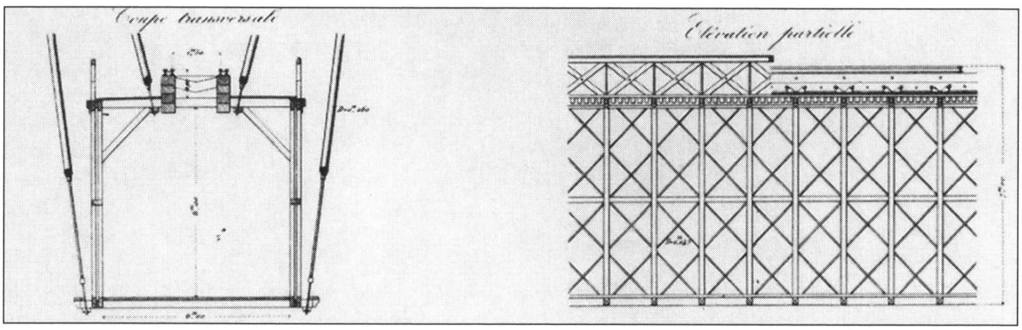

Roebling believed that this double-deck design was the only one possible at that time. Skittish horses could not coexist with puffing engines on the same level. As well, each of the decks, when bound together, would complement each other in providing the stiffness that the chief engineer insisted upon. This diagram shows a cross section of his design and the supporting latticework between the decks. (L. L. Buck, *Renewal of the Niagara Suspension Bridge*, 1880)

Roebling convinced his detractors that his bridge would stand up under moving trains because he would make it stiff and heavy enough. He would design the two floors and the open wooden trusswork, and then tie them together as one enormous hollow straight beam. The two decks would be connected by diagonal tension rods, making the structure stiff enough to support rolling trains. Their weight, however, would be supported by the main cables. The timber would be well seasoned and the upper deck caulked and painted. This would serve like the roof of an old-fashioned covered bridge. This woodcut is from the late 1850s. (*The English Cricketer's Trip to Canada and the United States*, 1859)

When completed, Roebling's bridge did appear to be a long narrow cage or box across the gorge. This photo from 1896 shows the bridge extending from the top of the bluff on the American side. The stone towers of the second bridge were replaced by steel in 1878 as part of engineer Leffert L. Buck's upgrading of Roebling's original towers, which, after two decades, were showing signs of deterioration. (NFBC Archives)

Work began on the stone towers in September 1852. Built well behind the towers of the first suspension bridge, they were constructed by stonemason John Brown on the Canadian side and the Lathan & Gage firm on the American side. With stone from the Queenston Quarries near St. Davids, Ontario, they became solid masonry structures 60 feet (18.3 m) high and 15 feet (4.6 m) square at the base. The decorative tops were capped with cast-iron plates over which the cables, in saddles, were strung. It took fourteen months to complete the towers. This 1870s photo gives an indication of the size of the towers in relation to the waiting carriage. (NF PL Heritage Foundation Collection)

This 1860s stereographic view of the town of Niagara Falls, NY, suggests a tranquil rural atmosphere linked to Canada by a spectacular bridge. The City of Niagara Falls was not incorporated until 1892, when it was in the process of becoming part of a highly industrialized area based on the power potential of the Niagara River. (John Burtniak Collection)

Anchor pits were dug for each cable, and the looped ends of each cable were fastened to chains of iron links. Then the whole anchorage was buried under a mass of concrete. This photo shows one of the sets of chains in one of the anchor pits exposed during their renewal in 1878. (NFBC Archives)

Stretching 821 feet (250 m) between the towers, the four giant cables were composed of seven bundles of wire, each made up of 520 No. 9 gauge wires. These were built up by feeding the wires into a travelling sheave that spun the wires into position. The process was a Roebling design and patent. The seven bundles were then bound into one cylindrical cable 10.5 inches (26.7 cm) in diameter. They were then wrapped for weather protection. The wire for these cables came from the Richard Johnson factory in Manchester, England, while the tons of other bridge wire were manufactured by the Roebling works in Trenton, New Jersey. This photo shows a close-up of the cables on the railway deck. (NF PL)

Two sets of suspender cables were attached to the main cables. One set was hung from the upper cable to the railway deck, and the other set hung from the lower cable to the carriage deck. In addition, large numbers of guy wires from the bridge to the cliff were employed to counteract the effects of the wind. This illustration shows both sets of suspension cables, the diagonal tension rods, and the guy wires. (NFBC Archives)

In this photo of the interior of the carriage deck, the extent of the wooden braces is clearly visible. The deck itself was covered first with a 2-inch (5 cm) layer of pine and then a 2-inch (5 cm) layer of oak planks. It was completed in April 1854, after which Ellet's old suspension bridge was no longer useful and was dismantled. (NFBC Archives)

The railway deck was completed late in 1854. The tracks were laid on pine beams, which were bolted to the beams on the roof of the carriage deck. There were four rails laid to accommodate three different railway gauges: the 5-foot, 6-inch gauge for the Great Western Railway system; the 4-foot, 9-inch New York Central system; and the 6-foot gauge for the Canandaigua & Niagara Falls Railway. This early photo shows an approaching engine on the narrow gauge, and the suspenders attached to the main upper suspension cable. (C. H. Ellis, *Railways*, 1974)

Each of the three railroads that used the bridge went their separate ways once across. This stereographic view shows the three lines emerging onto the American side. (John Burtniak Collection)

With bridge owners, engineers, and hundreds of spectators lining both sides of the gorge, the "London"—the largest locomotive belonging to the Great Western Railroad—steamed across the new bridge to the United States on March 8, 1855. It weighed twenty-three tons and pushed twenty double-loaded cars, which soon covered nearly the whole length of the rail deck. "No vibrations whatever," Roebling is reported as noting. This train became the first train to cross a suspension bridge in the history of the world. A week later, on March 15, the railway bridge was opened officially to rail traffic between the two nations. Its success overcame much of the skepticism about suspension bridges. Here, a Great Western train crosses over to the United States. (National Museum of Science & Technology, Smithsonian Institution)

While the engineers were assured of the bridge's safety, some people found that passage by carriage was a terrifying experience. Mark Twain's analysis of the crossing, for example, confirmed their fears. He wrote, "You drive over to Suspension Bridge and divide your misery between the chances of smashing down two hundred feet into the river below, and the chances of having a railway train overhead smashing down onto you. Either possibility is discomforting taken by itself, but mixed together, they amount in the aggregate to positive unhappiness." (Quoted in J. Gies, *Bridges and Men*, p. 188.) This early stereographic view peers into the lower deck at the tollgate. (John Burtniak Collection)

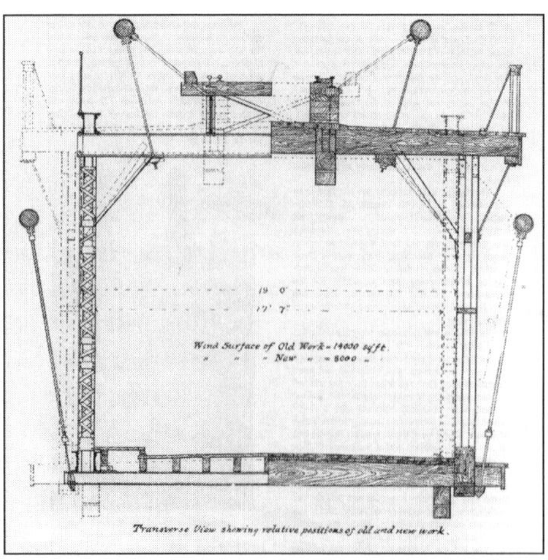

By 1860, an average of fifty trains a day passed from one country to the other. By the 1880s, the increased traffic and the creation of larger and heavier trains prompted the bridge owners to consider rebuilding the structure. This task fell to Leffert L. Buck, and after eight years of reconstruction, the masonry towers and the wooden framework were replaced by steel. Beauty was sacrificed for utility, and the bridge was able to serve for another decade until continuing increases in railroad weights made it necessary to replace the whole span with a totally new structure of more modern proportions (see Chapter Three). This diagram shows the bridge in wood on the right and steel on the left. (L. L. Buck, *Renewal of the Niagara Suspension Bridge, 1880)*

A railway train crossing a bridge certainly got people's attention, however nervous they may have felt. The sight of a moving train held up by such delicate-looking threads was, in the 1850s, nothing short of miraculous. Something so delicate should surely collapse under something so heavy. That it did not seemed pure magic. Roebling had proven that the principles of suspension could be applied safely. ("Radio Times," Hulton Picture Library)

When Buck completed his renovation work, the bridge was capable of supporting 350 tons of live load, doubling the capacity of the original 1855 structure. All of Buck's renovation work was done without any interruption of rail or pedestrian traffic. Roebling's original main suspension cables continued to serve, however, until they were taken down in 1897. This June 1897 photo of the new Steel Arch Bridge nearing completion shows one of those original cables already removed. When removed and cut, the pieces curled to the same "set" as they had when taken off their reels in 1853, indicating their undiminished strength. (NFBC Archives)

Tourist brochures of the 1860s through the 1890s highlighted the Suspension Bridge as one of the "must sees" of the area. As a significant highway of commerce and as a convenience to pedestrians, it fulfilled its mandate admirably. (John Burtniak Collection)

John Roebling had defied the doubts of his contemporaries and built a bridge that changed all preconceived notions. Leffert Buck, the builder of the succeeding Steel Arch Bridge in 1896 said of Roebling's bridge, "Even viewed in the light of increased experience and...despite flaws developed by long service, and reconstruction rendered necessary by time and abuse, the Niagara Railway Suspension Bridge will always be considered a monument to engineering skill." (NF PL Francis J. Petrie Collection)

Three

Niagara Railway Steel Arch Bridge

(Lower Steel Arch Bridge) (Whirlpool Bridge)

Above, the Whirlpool Rapids Bridge as it exists in the early 21st century. Its original name was the Niagara Railway Arch Bridge, and it was built by Leffert L. Buck in 1896–97. As soon as it opened, it became known as the Grand Trunk Railway Bridge. The popular name, however, became the Lower Steel Arch Bridge (shortly after the construction of the Upper Steel Arch Bridge a couple of years later). Its name changed yet again in 1937 to the Whirlpool Bridge, but it was known locally on both sides of the border as the Lower Bridge. (NFBC Archives)

Leffert L. Buck (1837–1909) Bridge Builder

Born and educated in Canton, New York, Leffert Buck attended both the newly established St. Lawrence University in his home town and, later, after an interruption in his education by service in the Civil War, the Rensselaer Polytechnic Institute at Troy, New York. He graduated as a civil engineer in 1868. A variety of engineering assignments carried him through the 1870s. His greatest enthusiasm, however, was in bridge design and construction, and by the 1880s he had become a living authority on suspension bridges in particular. His Williamsburg Bridge over the East River in New York City, which opened in 1903, is perhaps his most well-known and lasting engineering accomplishment. His reconstruction of Roebling's Niagara Railroad Suspension Bridge in the 1880s, however, was equally impressive. Here, over an eight-year period, the cable anchors were reinforced and the wooden trusses and stone towers were replaced with steel. All this work was done without interruption of traffic.

Eventually, increasing rail traffic and heavier locomotives pulling longer loads made complete replacement of the Roebling bridge necessary. Buck was called upon to design and build a new bridge to meet the needs of the new century. The chief engineer began his surveys of the banks of the river in the spring of 1893, but it was not until May of 1896 that the Niagara Falls International Bridge Companies approved his plans. It would be a steel arch, double railway track bridge. Buck again used his unique engineering skills to allow construction without interruption to rail traffic. Ground was broken for the foundations early in April 1896, and the contract for the superstructure was awarded to the Pennsylvania Steel Company in June.

In many ways Leffert Buck was the intellectual heir to John Roebling's pioneering legacy. His Whirlpool Rapids Bridge, known in his time as the Niagara Railway Arch Bridge, still stands as his own engineering legacy to the region. It opened in 1897 and celebrated its 100th anniversary a century later.

This rare 1898 photograph shows the executives of the Pencoyd Steel Company in the center flanked by Leffert L. Buck on the left and his resident engineer R. S. Buck (no relation) on the right. They are standing in front of their construction office. (NFBC Archives)

The bridge abutments (skewbacks) were dug deep into the Clifton limestone ledge, shown here in late August 1896. This rock layer first appears at the base of the Horseshoe Falls, but two miles farther downstream it is higher up on the gorge wall. The Roebling Railway Suspension Bridge juts out from above the gorge wall. (NFBC Archives)

The abutments (skewbacks) were under construction in late August 1896. These abutments, two on each side of the gorge, were built of gray limestone with granite copings. This stone was cut from the nearby Queenston quarries just north of St. Davids, Ontario. On the New York side, the abutments sat directly on the limestone ledge. On the Canadian side, however, concrete had to be used under the front of both. (NFBC Archives)

A heavy face wall under the New York skewbacks was needed because of the encroachment of the Gorge Road. The cut made for the road left an almost vertical face directly at the front of the skewbacks. The roadway appears here in front of the vertical face with the skewbacks high above. This photo was taken late in October 1896. (NFBC Archives)

This photo gives a close-up look at the skewbacks under construction in August 1896. The workers are dwarfed by these immense foundations built to support the arches to follow. The foundations were completed late in September 1896. (NFBC Archives)

Huge limestone blocks were cut to size above the gorge and then lowered by crane down to the skewback construction below. (NFBC Archives)

The steel towers continued to support the giant cables of the Roebling Suspension Bridge throughout the construction period. After the Steel Arch Bridge was completed, they were removed along with the framework of the Suspension Bridge. This view of the American side also shows the completed skewbacks. (NFBC Archives)

Erection of the bridge began in September 1896. After the skewbacks were in place, huge timber scaffolds (called *falsework*) were set onto the slopes and over the skewbacks on both sides of the gorge to support end spans and the first panels of the arch. Above is a view of the falsework on the American side taken late in September 1896. Below is a view of the falsework on the Canadian side taken in October. These views show how the engineers worked below and around the existing Suspension Bridge to allow for the uninterrupted flow of traffic. (NFBC Archives)

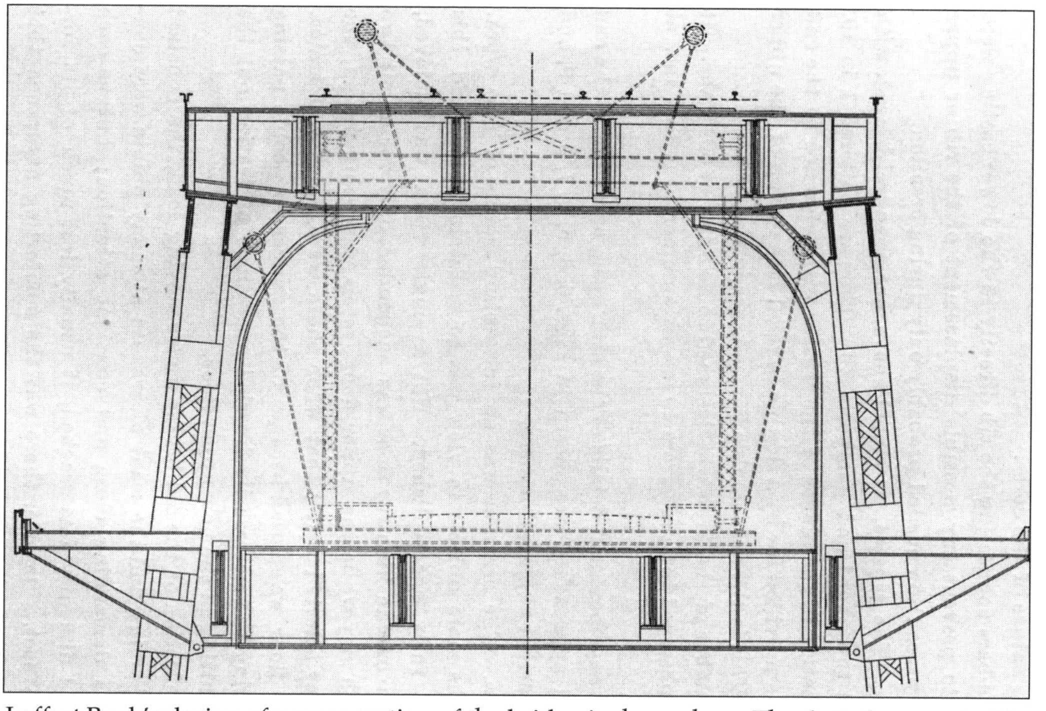

Leffert Buck's design of a cross section of the bridge is shown here. The dotted square inside the steel bridge structure shows how the Steel Arch Bridge was constructed neatly outside and around the Roebling Suspension Bridge. (L. L. Buck, *Report on the Construction of The Steel Arch Bridge*, 1899)

Steel travellers were built above the railway deck, one on each side of the gorge. These devices acted like stationary cranes and were designed to do the heavy lifting of the many sections of structural steel as the bridge was built outward from each bank. They were constructed high enough to clear the suspension cables and to give the necessary clearance for passing trains. The travellers rode on temporary tracks on each side of the bridge deck. (NFBC Archives)

Early in December 1896, the end posts (shoes) began to arrive on flatcars from the Pennsylvania foundry. (NFBC Archives)

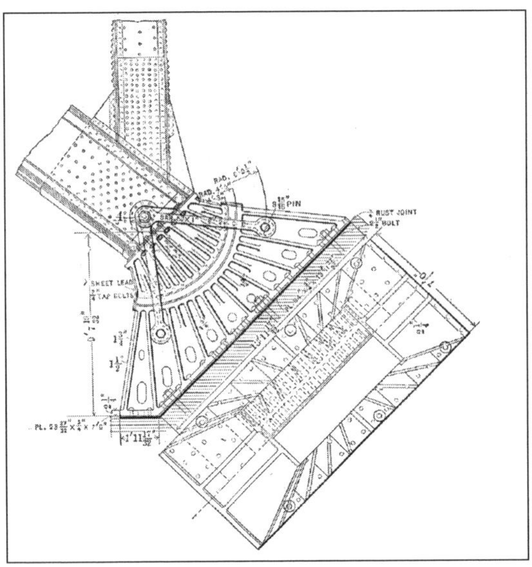

Shown to the left is Leffert Buck's diagram of an end post connection with the first rib. The end posts sat on a 2-inch-thick pad called the *rust joint*, made up mostly of cast-iron filings glued together. The end posts were then bolted into the skewback. (L. L. Buck, *Report on the Construction of The Steel Arch Bridge*, 1899)

The installation of the end posts always occasioned tense moments. While the engineers' measurements were thoroughly checked and rechecked, it was only when the end posts were settled perfectly onto their skewback bases that everyone could breathe more easily. The first end post was bolted into place on its skewback on the Canadian side on December 14, 1896. The others quickly followed. (NFBC Archives)

Completing the installation of the end posts always offered a photo opportunity. This shot was taken on January 13, 1897. (NFBC Archives)

The steel panels and huge rib pieces began arriving at both bridgehead sites early in January. (NFBC Archives)

With the end post castings in place, Chief Engineer Leffert Buck and his resident engineer Richard Buck were ready to begin the lifting of the first panels of the arch down to the end posts. This beginning of the actual arch construction was accomplished by the end of January 1897. Construction proceeded simultaneously from each side of the cliff. The traveller on the Canadian side in this photo has just lifted the first rib sections into place. When joined together, these rib sections were supported by cables—one end fastened to the ends of the arch section and the other end slung over the towers of the old Suspension Bridge and attached to its existing anchors. (NFBC Archives)

With the first panels of the arch in place and supported by the falsework, the two halves of the arch were built out as cantilevers anchored to the solid rock on top of each cliff. (See Chapter Four for the cantilever principle.) The two travelling cranes carried each precisely measured steel member out to position. This photo from mid-February 1897 shows the early progress of the arch. Spectators could watch the construction from relative safety inside the old Suspension Bridge. (NFBC Archives)

By mid-March, the progress was more dramatic. As the arch was completed panel by panel, the travellers moved forward, approaching the midway joining point. (NFBC Archives)

This photo from March 16, 1897, shows the two sides approaching each other. Behind the arch is the Cantilever Bridge belonging to the Niagara River Bridge Company (Cornelius Vanderbilt, President). It was completed in 1883 and was strictly a railway crossing for the Canada Southern and Michigan Central Railway Companies (see Chapter Four). (NFBC Archives)

Early in April 1897, the two sides of the arch were allowed to come together. After some initial difficulties, the closure was completed and fastened with rivets. The entire bridge structure then became rigid. At a total length of 1069 feet (326 m), it was the longest arch bridge in the world to that date. From abutment to abutment, it spanned 550 feet (168 m). It represented a triumph of engineering for Leffert Buck. The structure is the only 19th-century bridge spanning the Niagara Gorge that has survived into the 21st century essentially unaltered. (NFBC Archives)

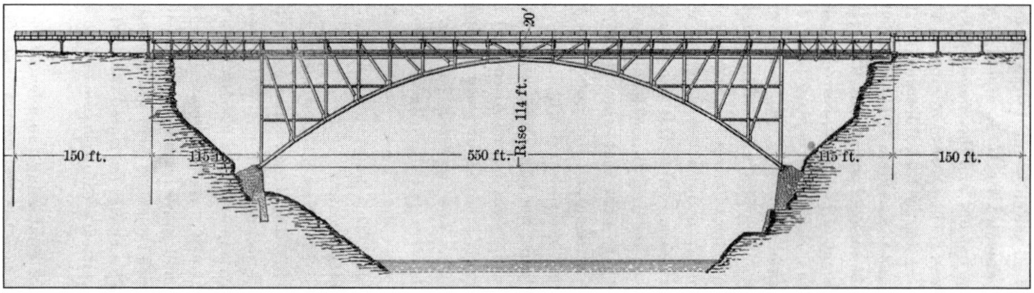

Early in the design stage, Leffert Buck created this diagram of a cross section of the bridge. The Canadian side is on the left. The American side, showing the ledge for the Niagara Gorge Railway roadway, is on the right. (L. L. Buck, *Report on the Construction of The Steel Arch Bridge*, 1899.)

In May and June of 1897, the old suspension bridge was taken apart, and the giant cables were lifted off their towers by hydraulic jacks. The material, including the dismantled travelling cranes, was sent to the Pennsylvania Steel Company for recycling. The pedestrian way under the rail deck was planked, and the rails on the rail deck above were replaced. The bridge was tested late in July, and all work was completed on August 27. This photo was taken on July 31, 1897, just eight months after the skewback shoes were set in place. Cleanup operations were nearing completion. (NFBC Archives)

This photo *circa* 1907 shows the bridge fully loaded with trains moving in both directions. (*Niagara in Summer and Winter*, ca. 1907)

This photo shows the replanked lower deck of the completed Railway Arch Bridge. The only suspension of traffic on this floor (for safety reasons) took place during a two-hour period each day when the upper floor carrying the new double rail track was being installed. (NFBC Archives)

The photo above shows a similar scene taken in 1988. The original double track on the top deck was reduced to a single track in 1963 as the importance of rail transportation declined. In 1967 the wooden floor on the traffic deck was replaced and topped with asphalt. (NFBC Archives)

This photograph of the bridgehead on the American side was taken in June 1897. Here, three rail lines converge to become two on the bridge's rail deck. (NFBC Archives)

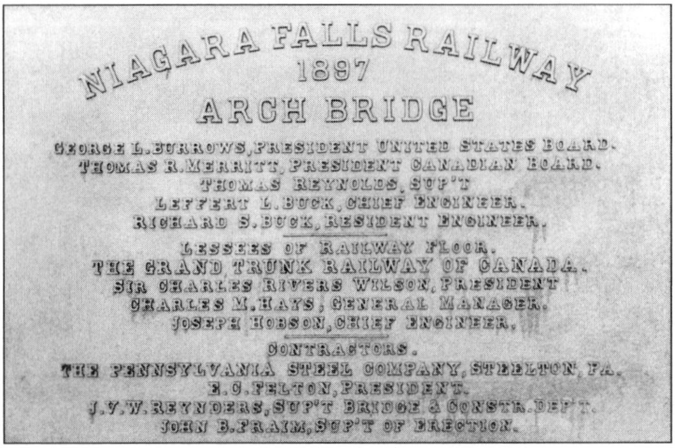

The official bridge opening celebrations took place over a three-day period from September 23–25, 1897. A carnival was being held at that time, and the bridge festivities blended in nicely. Band concerts, variety performances of all kinds, shows, dime museum exhibits, and beer gardens catered to the visitors. Special trains from nearby cities brought the throngs to fill the hotels to overflowing on both sides of the border. The Mayor of Clifton even declared a half-holiday beginning at one p.m. on Friday afternoon. That evening—with 150 bandsmen from Canada and the United States playing the national anthems, and twenty-one aerial bombs fired from the center of the rail deck—the bridge was declared open. The plaque, shown above, was unveiled by George L. Burrows and Thomas R. Merritt, the presidents respectively of the American and Canadian Boards of Directors of the Niagara Railway Arch Bridge. (NFBC Archives)

One writer predicted in *Engineering News* in 1898 that "there is every reason to anticipate that the structure will be capable of serving the purpose...a hundred years from now as it is today." This belief was fulfilled in two days of centenary celebrations in September 1997. A host of dignitaries, including the Lieutenant Governor of Ontario, Hilary M. Weston, were on hand on Friday the 12th to rededicate the bridge and to unveil a commissioned painting by artist Paul Hanover. The bridge was actually closed both days for the first time in one hundred years to allow the general public to enjoy a stroll and a mix of picnics, bands, dancers, and singers. The celebrations, including spectacular fireworks and an enormous cake, were sponsored by the Niagara Falls Bridge Commission, the agency that had owned the bridge since January 1959. The logo, shown above, was created for the event. In 2004, with the increasing need for border security and with improved facilities at the nearby Rainbow Bridge, the Whirlpool Bridge became a crossing point for NEXUS users only.

Four

The Niagara Cantilever Bridge
(Grand Trunk Railway Bridge)
(Michigan Central Cantilever Bridge)

Designed by C. C. Schneider, one of the first modern steel cantilever bridges in North America became operational in 1883. Using methods that would be used in the construction of future cantilever bridges, Schneider's Niagara Cantilever Bridge was erected at a gorge site a few meters south of the Niagara Railway Arch Bridge. The Central Bridge Works of Buffalo was the chosen builder on behalf of the Niagara River Bridge Company. Late in 1882 they chose Schneider as their chief engineer. Schneider's utilitarian bridge would have a useful life of over forty years during a time when railroad traffic was flourishing. The speed of its construction in 1883, coupled with its efficient performance from its beginning, made it one of the most innovative bridges in the world at that time. It can claim to be the world's first modern cantilever bridge. The picture above shows the cantilever bridge with the Upper Suspension Bridge in the distance. Many area guidebooks of the late 19th century used this etching to show the relative positions of the bridges. (*Tugby's Illustrated Guide to Niagara Falls,* 1889)

Charles Conrad Schneider (1843–1916) Bridge Builder

Charles Schneider was born in Saxony, Germany, in 1843. He was educated there, and he trained and practiced as a mechanical engineer before coming to the United States in 1867. He began work, as many immigrant engineers did then, as a draftsman. His early work with the Rogers Locomotive Works in Paterson, New Jersey, led to his involvement with railroad companies, and before long he was in charge of all engineers for the Erie Railroad. It is understandable that bridges came to be designed in railroad offices, as the rail companies continually faced the need to bridge difficult topography. With the Erie Railroad, Schneider came to have responsibility for checking the bridge plans submitted by various bridge companies. In particular, he was responsible for checking the strain sheets, which showed the amount of pressure that each part of the bridge was designed to sustain. It was a natural progression for Schneider to become involved in actual design. In 1878, he established his own engineering firm in New York City. One of his first commissions was a series of bridges for the Canadian Pacific Railroad, including a proposed cantilever bridge crossing the Fraser River in British Columbia.

In the 1870s, the engineering profession became enamoured with the cantilever principle as a method of building stronger bridges, especially where the use of falsework to support the early stages of construction was not feasible. Schneider chose a cantilever design as best suited to the Niagara Gorge location. Two giant arms, set on stone piers, would support the center span. His cantilever bridge at Niagara, and the nearly identical Fraser River span built almost at the same time, directed the attention of the engineering world to this new type of bridge. They became prototypes for the subsequent cantilevers at Poughkeepsie, NY; the Firth of Forth Bridge in Scotland; and the Quebec City Bridge.

Schneider's Niagara bridge was as significant an engineering achievement as Roebling's Railway Suspension Bridge. In the 1890s, Schneider was the chief engineer for the Pencoyd Bridge Company, which built the superstructure of the Upper Steel Arch Bridge that opened in 1898. In 1905, Schneider was awarded the Norman Medal of the American Society of Civil Engineers for his contribution to engineering science.

In an 1887 lecture given by British civil engineer Benjamin Baker at the Royal Institution in London, the cantilever principle was explained by a demonstration. Baker used three men as a living model to show his general audience the nature of the stresses on this type of bridge. Two men on chairs supported sticks butting against the chairs. These represented the piers of the cantilevers. On the central girder sat a Japanese engineering student supported entirely by the counterweights represented by two piles of bricks. When stresses were brought onto the system by a load on the central girder (the Japanese engineer), the men's arms and the anchor cables came into tension, and the sticks and chair legs came into compression, thus supporting the central girder. The photo shown above became famous and was reproduced in many journals, but the cantilever principle was well understood by C. C. Schneider a decade earlier.

The Eads Bridge at St. Louis, built in 1874 and the first to use steel as opposed to wrought iron, was one of the earliest examples of the use of the cantilever principle in America. The famous Firth of Forth Bridge in Scotland, begun in 1883, is another early example.

In the early 1880s, the Michigan Central Railroad was looking for a rail link between Canada and the United States. Not prepared to pay the high rental cost that the owners of the Lower Arch Bridge were asking, the Michigan Central and its affiliate, the Canada Southern, decided to build a separate bridge. The Vanderbilt family (which controlled both systems) formed the Niagara River Bridge Company and negotiated the necessary government charters. A construction contract with the Central Bridge Works Company of Buffalo was signed, and surveys began in the fall of 1882. Ground was broken for construction on April 15, 1883. This photo from the early 20th century shows the completed bridge with the Lower Arch Bridge to the north. (*Niagara in Summer and Winter,* ca. 1907)

Actual work on the bridge began later in April 1883. Deep holes for the foundations were excavated by blasting into the shale. These were then filled with a new product called Boston Cement, said to be stronger than concrete. They were ready to receive the stone piers (abutments) by late June. These masonry piers were built of Queenston limestone quarried near St. Davids, Ontario. The stone blocks were lowered from the top of the gorge by derrick to the river's edge where mules dragged them to the pier site. When completed in August, each one measured 11 by 37 feet (3.4 by 11.3 m). This later photo, with two trains crossing, shows how close to the river's edge the abutments were placed. (*A New Guide to Niagara Falls,* Sherman Zavitz Collection)

Work on the steel towers began immediately after the piers were in place. Again, derricks lifted the steel members down to the pier site. When completed in early October 1883, they stood 132 feet (40 m) tall, high above the rapids and ready to receive the two arms of the cantilever. The shore arm was supported at first by falsework and then anchored firmly to the rock at the top of the gorge. This allowed the outer or river arm to be extended out over the water a distance of 175 feet (53 m) beyond the towers. The shore arms were balanced by the river arms, which were erected piece by piece, by means of travelling derricks. (John Burtniak Collection)

Working against a deadline imposed by the Vanderbilt rail interests, the steel sections advanced rapidly some 25 feet (7.6 m) every two days. The remaining gap of 120 feet (37 m) was filled by an ordinary truss type section hung from the ends of the river arms. Schneider incorporated the ingenious use of an expansion joint between the cantilevers and the central span. This last central section was set in place in a matter of five days, a remarkable feat for its time. The double track was soon laid, and work was completed late in November 1883. This photo was taken shortly after completion. The Roebling suspension bridge can be seen in the distance through the shore arm of the cantilever. (NFBC Archives)

This photo from late 1896 shows the Cantilever Bridge just behind Roebling's Railway Suspension Bridge. The travellers are in place, marking the beginning of construction of the Niagara Railway Steel Arch Bridge. When completed in 1897, the Railway Arch was to extend 780 feet (238 m) cliff to cliff, while the Cantilever, when completed in 1883, extended 906 feet (276 m) across the gorge. Its double track had the capacity to bear the weight of two trains crossing at the same time. (NFBC Archives)

The first test train pulled out onto the Niagara Cantilever Bridge on December 6, 1883. It was only an engine and passenger car holding a variety of railroad officials. A second crossing was made by the newspapermen and other invited guests, followed by a third using a heavier freight engine. The bridge was then declared safe. Its formal opening was slated for December 20. The *Illustrated London News* showed this woodcut of the ten locomotives chugging across the bridge from the Canadian side, presumably at the formal opening. In the distance is the Upper Suspension Bridge. (*Illustrated London News*, January 1884)

The *New York Times* sent a correspondent to cover the bridge opening, truly an exciting event in the eyes of witnesses. The correspondent reported the arrival of excursion trains from nearby towns and crowds of spectators lining both banks of the river, packing the nearby Suspension Bridge for a better view.

A heavy snowfall greeted the day of the final test. He wrote, "The test trains bedecked with flags started side by side from the Canadian bank. They consisted of 10 locomotives and 11 loaded gravel cars on each track, with an estimated weight of 1900 tons. They stopped on each arm of the cantilevers and also on the middle truss in order to make efficient and thorough tests....The trains were then backed to the Canadian side and in a few minutes a procession of visitors and townspeople, headed by the Tuscarora Band, proudly marched on the broad foot walks to the Canadian shore."

While the local folk marvelled at the structure, the engineers huddled with their test results. They reported that after the trains reached the middle of the bridge, the maximum deflection from the ends was just over six inches. When the trains left the bridge, all the parts returned to their original position—exactly as had been predicted. A grand banquet at the Monteagle Hotel on the American side followed the celebrations.

At the banquet, one local newspaper editor was quoted as saying, "We think it is peculiarly fitting that this latest triumph of engineering skill—like the suspension bridge—should be first demonstrated in view of the world's greatest wonder [Niagara Falls]. There let them stand, side by side, as proud monuments of nature's and man's most perfect and awe-inspiring handiwork."

A simple plaque was attached to the abutment on the American side. Along with the names of the president, the manager, the treasurer, and the resident engineer, it read: "Work commenced April 19, 1883. Work finished December 1, 1883." Strangely, while the chief engineer's name is on the invitation to the official opening, it does not appear on the plaque.

Below is the official invitation to the opening, December 20, 1883. (Buffalo and Erie County Historical Society)

As heavier locomotives came on stream in the early years of the 20th century, plans were set in motion to replace the bridge. Because of the interruption of the First World War and the shortage of steel, it was not until 1923 that the Michigan Central Railway was ready to proceed with another steel arch. The replacement was opened to train traffic in February 1925 (see Chapter Five). Shortly thereafter, work was begun to dismantle the Cantilever Bridge. As the dismantling began, it became apparent that significant deterioration of the girders had occurred, thus eliminating the potential for recycling the structure for another location. The framework was then simply cut up for scrap, and all evidence of C. C. Schneider's Cantilever Bridge disappeared.

Five

Michigan Central Railway Arch Bridge

Of all the Niagara Gorge bridges, the Michigan Central Railway Arch Bridge drew the least attention from the general public. It was strictly a utilitarian structure, designed to cope with the ever-increasing weight of trains and train traffic in general after the turn of the century. The Michigan Central made plans to construct a replacement for its Cantilever Bridge during the First World War, but that conflict delayed construction until steel became available again in the post-war period. It was not until 1922 that their plans went forward. Work on the foundations began in April 1923 and was completed in October. Structural work then followed and continued through 1924. In January 1925, work crews realigned the tracks from the Cantilever Bridge approaches, and the bridge was opened with no fanfare on February 21. (NF NY PL)

Each arch was flanked by an approach span, 100 feet (30 m) at the Canadian end and 123 feet (37 m) at the New York end. To compensate for variations in temperature, the necessary expansion joints were placed in these approach spans and not in the arches. The arches themselves were built out from each shore, firmly attached to their approach spans. Each half of the arch was held back by two backstays consisting of steel eyebar links that were carried back into two chambers on each side. These were tunnelled deep into the cliffs. Heavy steel girders were placed in these chambers to which the backstay links were connected. The chambers were then concreted in to hold the arch as it was constructed outward. This photo from 1924 shows the arch attached to its approach span, far left. Underneath, the two backstays are visible. A derrick on the bridge deck, sitting on its steam-operated traveller, lifts the steel members out to the working end. The Cantilever Bridge, shown behind the new arch, continued to carry freight throughout 1924. (NF NY PL)

The bridge was designed by Hans Ibsen, a civil engineer who spent his entire career with the Michigan Central constructing a variety of bridges and supervising the maintenance and strengthening of many more. Its main arch would stretch 640 feet (195 m), supporting a double rail track, low guardrails, and a narrow walkway along one side. This aerial photo from late in 1925 shows the Cantilever Bridge in the foreground being dismantled alongside the newly operational Michigan Central Railway Arch Bridge. Ibsen died in his native Norway in 1940. (NFBC Archives)

Under each backstay, large 3000-ton capacity jacks were arranged so that they would either lengthen or shorten the backstays by the few necessary inches to bring the two huge halves of the bridge exactly together to complete the arch. After the arch became self-supporting, the backstays between the ends of the arch and the cliff face were removed, but the steelwork in the anchorage chambers and the mouths of the chambers themselves were filled in with concrete and smoothed over. This photo from 1925 shows the two arches coming together. The Cantilever Bridge is behind it. The Niagara Railway Steel Arch Bridge is in the foreground. (NFBC Archives)

This photo from late November 1924 shows the two arches approaching one another. The sections were run out onto the arch where the huge cranes lifted them out and down to the working ends. (NF NY PL)

As with most bridge constructions, the closing of the arch (making it self-supporting) was a chance for a photo opportunity. This event at the new Michigan Central Bridge site took place in January 1925. The freight cars in the background are sitting on the Cantilever Bridge. The work crew numbered just over seventy on this project. (NFBC Archives)

This 1925 view of the completed bridge was one of the last taken from the Cantilever Bridge as it was being dismantled. The Whirlpool Bridge is seen behind it through the girders. (NFBC Archives)

Reflecting the ever-changing nature of railroad ownership, the 1925 Michigan Central Railway Arch Bridge eventually became part of the New York Central system, and then part of the Penn Rail system, and later still by the Consolidated Rail Corporation. In 1984, joint ownership between Canadian National and Canadian Pacific was arranged when they purchased the assets of the Canada Southern Railway Company. Canadian Pacific operated the bridge into the 1990s, but all operations ceased in December 2001. The remaining single track was soon torn up, and a security barrier was set in place. This photo from 1990 shows the single track still in use. (NFBC Archives)

This photo from 2002 shows the track area closed off with a barbed-wire fence. The single track was still intact. (NF NY PL)

The old Michigan Central Railway Arch Bridge no longer serves its original purpose and has been abandoned. Significant declines in rail traffic in the late 20th century spelled the end of its usefulness. There have been proposals for redevelopment, but as of 2008, the bridge was for sale. This photo from 2008 shows a reinforced security barrier blocking access to the rail-less superstructure. (NFBC Archives)

Six

Upper Suspension Bridge

(New Suspension Bridge)
(Falls View Suspension Bridge)
(Niagara-Clifton Bridge)

The success that Roebling's Suspension Bridge enjoyed in the 1850s and 1860s created agitation for a crossing closer to the Falls. Tourists looked for proximity while the local businesses longed for convenience. It was not until 1867, however, that the Niagara Falls Suspension Bridge Company (US) and the Clifton Suspension Bridge Company (Canada) received their charters to span the river with another bridge. In June 1867, the directors chose Samuel Keefer of Thorold as their chief engineer. Work began almost immediately to clear the site in preparation for actual construction early in 1868. The bridge would be completed in less than a year. This engraving shows Keefer's bridge as it stood in the late 1870s. (*The Falls of Niagara,* ca. 1880)

Samuel Keefer (1811–1890) Bridge Builder

Samuel Keefer was born in Thorold, Ontario, in 1811. He was educated there and later at Upper Canada College in Toronto. He began his life's work as an apprentice with the Welland Canal Company in 1827, most likely as a labourer. This project to link Lake Erie with Lake Ontario to get around the barrier at Niagara Falls had begun in 1824 and was an ideal preparation for an engineering career. In 1833 he became the secretary of the Board of Canal Commissioners, which was investigating improvements to shipping on the St. Lawrence River. In 1834, Keefer became the assistant to the chief engineer of the new Cornwall Canal, a position he held for five years. In 1839, he was appointed as the first chief engineer to the newly created Board of Works, which became the Department of Public Works of the United Provinces in 1846. It was the highest engineering position in Canada at that time. In the 1840s, this government agency was responsible for the completion of the St. Lawrence canal system. In the 1850s, it was responsible for the planning and construction of the provincial railway system.

In this position, Keefer had the opportunity to study and practice bridge design and bridge building. One of his first designs was Canada's first suspension bridge across the Chaudière Falls between Ottawa, Ontario, and Hull, Quebec. It was completed in 1843. Keefer retired from public service in 1864 but not from his profession. In 1867, he undertook the design and construction of the Clifton Bridge over the Niagara River. When completed early in 1869, with its span of 1268 feet (386 meters), it was the longest such structure in existence in the world. For this, he ultimately won a gold medal in the engineering competition at the 1878 Universal Exposition in Paris. His Clifton Bridge was the fourth to span the Niagara River. Interestingly, he had submitted designs for the first Niagara bridge in 1846, but the directors awarded that contract to Charles Ellet. Keefer's later years were spent in semi-retirement. He joined the American Society of Civil Engineers in 1869, and in 1887 he helped found the Canadian Society of Civil Engineers. His dedication to his profession did much to elevate civil engineering in Canada to professional status.

Keefer's four towers were completed in the fall of 1867. These were made of 12- by 12-inch (30 by 30 cm) pine timbers with bases 28 feet (8.5 m) square. Each leg sat on an iron shoe. With the rock base slightly different on each side of the gorge, the towers on the American side stood 100 feet (30 m) high while the Canadian towers reached 105 feet (32 m). A saddle plate for the main cables topped each one. With construction virtually complete early in 1869, this photo shows the towers, cross braced and bolted together. (NF NY PL)

The first 2-inch (5 cm) cable was dragged across the ice bridge in February 1868. It would serve as a carrier wire on which to pull larger cables across. Two larger wires were then strung across the top of each tower and anchored to serve as working cables. Two platforms were then built to run along each of these, from which the workmen could work on the main cables. This early summer 1868 photo shows chief engineer Keefer himself using one of these platforms to ferry the visiting Chinese Ambassador. (NF NY PL)

The anchorages to accept the ends of the cables were dug late in 1867 and completed in the spring of 1868. On the American side, they were embedded in masonry 18 feet (5.5 m) below the surface. On the Canadian side, they were quarried out of solid rock. This photo shows the anchorage on the Canadian side, which was exposed in 1884–85 when the wooden towers were being replaced. (Buffalo and Erie County Historical Society)

The main cables made of English iron were brought over the towers in the summer of 1868. Between the anchorages they stretched 1820 feet (555 m). Each consisted of seven strands of No. 9 wire, all bound together. This photo from midsummer 1868 shows the wire floor stays (suspenders) already attached to the main cables at 5-foot (1.5 m) intervals. (NF NY PL)

The roadway would be built out from each side of the gorge. First, stringers 10 feet (3 m) apart would extend. Cross ties 5 feet (1.5 m) apart would join them together. A double floor of Norway pine then joined the cross ties. This photo from the late summer of 1868 shows the right and left stringers, the cross ties joined to the main cables, and the first layer of floor planks. (NF NY PL)

This photo shows the bridge deck extending out from the American side and nearing the halfway point. The roadway planking was about half complete. The two sides would meet in mid-October. The wire suspenders decrease in length toward the bridge's center, thus keeping the roadway level. (NF NY PL)

The bridge was strengthened by iron bars running the whole length of the underside of the roadway. Stiff side railings 5 feet (1.5 m) high were bolted to the floor beams. The flooring was completed late in December 1868, and local residents were invited to inspect the work for free. Keefer's masterpiece opened for traffic on January 2, 1869, only slightly behind schedule, and it passed inspection later in the month. It was 10 feet (3 m) wide with a raised sidewalk. Carriages moved across in one direction only, alternating from each side at fifteen-minute intervals. This photo from the mid-1870s shows the system of suspenders and the raised sidewalk for the best viewing of the Falls, out of the way of carriage wheels. (NF NY PL)

From its opening in January 1869, the bridge was a financial success. Local businesses now had the convenience that they had longed for, and the tourists were thrilled to be so close to the grandeur of the Falls. One 1880s travel book proclaimed that the bridge's "great length, symmetrical form, graceful curve, and obvious strength, cannot fail to strike every beholder with equal surprise and pleasure....It is of itself an object of curious interest...for the inspection and study of the great natural wonders displayed here in such magnificence." This photo is looking at the Canadian side. (NF NY PL)

In 1872, the Bridge Company decided to upgrade the entrances on both sides. Two towers were built and covered with wood on three sides. The landward sides were covered with corrugated iron as fire protection. Rooms for customs officials and bridge directors were created under the towers, and an Observation Deck topped each one. For ten cents, tourists could ride to the top in an Otis steam-powered elevator on the Canadian side. This mid-1870s photo shows the new entranceway on the Canadian side as well as the system of guy wires attached to the bluffs at the time of construction to stabilize the entire structure. A maintenance crew is at work. (Sherman Zavitz Collection)

Each tower was topped with a glass-enclosed Observation Deck. Above that, a set of stairs climbed to an open-air Promenade Deck. While the elevator on the Canadian side was an added attraction, it did not prove to be a financial success for its operators. This view is of the American entrance looking across to Canada. (NF NY PL)

In 1876, the wooden deck supports (cross ties) were replaced with iron. In 1881, the bridge deck itself was replanked. In 1884, the directors decided upon a total upgrade. The Elevator Towers were removed, and the wooden tower frameworks were replaced with steel. This construction photo from 1884 shows the new steel tower encasing its wooden predecessor. (NFBC Archives)

There had long been frustration over the bridge's narrow roadway. Ensuing traffic congestion led to an 1886 decision to rebuild everything except the new steel towers. The Rochester Bridge Works was contracted to replace all the wood, to supply all new wire suspender cables, to add two new main cables, to renovate the anchorages, and to widen the roadway. All this was completed in the fall of 1888. The newly widened Upper Suspension Bridge now had a 17-foot (5.2 m) roadway allowing two-way carriage traffic. This late 1887 photo at the Canadian Customs office shows the new steel entranceway, the new twin cables, and two carriages halfway across the bridge. The two additional main cables consisted of seven twisted galvanized wire ropes attached to eyebars anchored in the rock behind the towers. (NF NY PL)

The 1888 reconstruction led to the substitution of a lateral wind cable replacing the network of river guy wires that had performed admirably since 1869. New guy wires were anchored at the bottom of the gorge to prevent uplift by wind under the floor. All of this would not be enough. An unprecedented winter gale on the night of January 9, 1889, brought the deficiencies to everyone's attention. The bridge collapsed. All that was left the next day were the undamaged steel towers and main cables. The remains of the bridge lay on the slopes below and on the river bottom. Fortunately, no lives were lost. (NF NY PL)

Undaunted by the disaster, the directors signed a new contract with the Rochester Bridge and Iron Works to rebuild the bridge immediately. The plans from the 1887–88 rebuilding project were revisited, and construction was completed by May 1889, just four months after the former bridge's destruction. This project was led by chief engineer George W. McNulty (1850–1924). McNulty, a graduate of the University of Virginia, had been an assistant to the Roeblings as they built the famous Brooklyn Bridge, which opened in 1883. McNulty made sure that a new system of storm cables attached to the new bridge was included. These can be seen in the photo above, attached between the bridge and the bluffs. (NF NY PL)

This new suspension bridge served pedestrians and light carriages for eight years. During that time, road traffic increased, and there developed a need to accommodate the electric trolley line running along the top of the Niagara Gorge to complete a belt line link to the United States. For those reasons, the directors decided in 1894 that an entirely new steel arch bridge should replace the old suspension bridge. Construction for it began in 1897 under the direction of chief engineer L. L. Buck. The Upper Steel Arch Bridge was opened to traffic in June 1898 (see Chapter Seven). This photo from October 1897 shows the falsework erected under the Upper Suspension Bridge in preparation for the new bridge construction. The cables from the old suspension bridge would be removed in late spring of 1898 and taken to the site for the new Lewiston-Queenston suspension bridge (see Chapter Ten). (NFBC Archives)

Seven

Upper Steel Arch Bridge
(International Bridge)
(Falls View Bridge)
(Honeymoon Bridge)

In January 1897, the decision was made to build an entirely new structure to replace the Upper Suspension Bridge (built in 1869 by Samuel Keefer; remodelled and widened in 1887–88; and rebuilt again under chief engineer G. W. McNulty in 1888). Leffert L. Buck was named the chief designer and engineer. He and his resident engineer Richard S. Buck (no relation) were already at work on the designs and plans. Leffert Buck was a logical choice since he was completing the Niagara Railway Steel Arch Bridge at the time (see Chapter Three). The Suspension Bridge owners had employed Buck earlier to do a feasibility study on enlarging the old bridge to accept trolley tracks. This study had concluded that only a new steel arch would meet the demands both of the electric trolley line and of the ever-increasing commercial and pedestrian traffic. The Pencoyd Bridge Company of Philadelphia was awarded the contract, and the new Upper Steel Arch Bridge was completed for its owner, the International Traction Company, in just over a year, opening in June 1898. This view of the newly completed steel arch is from one of the many postcards generated in the early 20th century. The framework has been highlighted in white. (John Burtniak Collection)

The abutments (skewbacks) for the new arch bridge were completed early in 1897. They were 67 feet (20.4 m) apart. On the Canadian side, they sat on a base of rocks to facilitate access. On the American side, a rock platform was not possible, so the abutments were anchored to the native rock formation that existed at that level. This photo of October 1897 shows the limestone abutments on the Canadian side, close to the river's edge. (NFBC Archives)

The proximity of a strong outflow tunnel from the Niagara Falls Power Company made it necessary to shift the axis of the new bridge at the American end about 14 feet (4.3 m) upstream. This meant that the Upper Steel Arch Bridge would not be directly under the old Suspension Bridge on the American side. This photo of the construction in April shows the steelwork on the U.S. side slightly south of the existing Suspension Bridge rather than directly under it. The cables of the old Suspension Bridge continued to serve their original purpose until June 1898, when they were removed. (NFBC Archives)

The wooden falsework began to appear in the summer of 1897. The Pencoyd Iron Works and Bridge Company bought the timbers that had already been used in the construction of the Railway Steel Arch Bridge. The dismantling at that site took place in the spring and summer of 1897, just in time to be reassembled under the Suspension Bridge to support the shore arms of the new steel arch. These photos from mid-October show the wooden timbers nearing the height of the Suspension Bridge on the American side (above) and on the Canadian side (below). The abutments on the Canadian side are seen sitting on a base of rocks, while on the American side they are anchored to the native rock layer. (NFBC Archives)

This falsework was being used for the second time, effectively recycling it from the Niagara Steel Arch Bridge construction. While it looks rather fragile, it was securely bolted together and anchored to the cliffs behind. (NFBC Archives)

The two bow-shaped shore arms were almost complete by the end of October 1897. Work proceeded from both sides of the gorge at roughly the same pace. (NFBC Archives)

The entranceways to the Suspension Bridge were kept relatively clear for pedestrian traffic during the construction period, although as this photo of October shows, the approaches were somewhat cluttered. Traffic had to be suspended for only a three-week period in mid-April. The anchor chains of the Suspension Bridge in the left and right foreground were exposed at this time for the adjustment of the main cables. (NFBC Archives)

One of the challenges of bridge construction was the planning for the precise timing of the arrival of the various parts. This planning was all a part of the engineer's work. Steel from the Philadelphia foundries began arriving in January 1898. Late in the month, the four giant shoes arrived on flatcars for installation on the abutments. These would be lowered by crane to the river's edge. (NFBC Archives)

This February 1898 photo on the Canadian side shows the abutment, the shoe, and the roller bearing in its hinge bolt. The workers are obviously proud of their installation work. Resident engineer Richard Buck stands on the beam at the top right. (NFBC Archives)

Installation of the shoes onto the abutments was completed in early February 1898. As with the recently completed Railway Steel Arch Bridge, photo opportunities were taken at significant stages of construction. Onto each shoe was set a huge hinge, or anchor bolt, on which the arch revolved. These would accept the first steel sections. This photo shows the abutment, the shoe, and the roller bearing in place. Inspecting the work, top right, is the chief engineer Leffert L. Buck himself. (NFBC Archives)

The first section of steel for the arch was installed around February 10th. The photo shows the rock base in front of the abutments. Two ladies, quite probably the wives of Leffert Buck and Richard Buck, are taking advantage of the occasion with this photo opportunity. (NFBC Archives)

Above the first section of steel for the arch rose the first section of framework that would hold up the roadway bed. These vertical sections would rest on the arch itself. They decreased in height as they approached the center. The sections of the arch were held in place by cables attached to the Suspension Bridge. This photo was taken on March 23, 1898. (NFBC Archives)

The heavy lifting was done by mobile derricks called *travellers*. These ran along the Suspension Bridge railing, which served as a track for the traveller legs. This photo shows the traveller at work on the Canadian side, lowering a steel girder. The legs of the traveller were resting on the railing of the Suspension Bridge, and the girders rode out to the work area on the little truck shown here on its rails. (NFBC Archives)

The precisely measured steel girders were stored close at hand just south of the entranceway to the Suspension Bridge. Using the block-and-tackle method, these would be levered onto the small buggy or truck on rails. That would in turn ferry them out to the end of the work area where the traveller would lift and then lower them into place for riveting. (NFBC Archives)

By early April 1898, each arch was approaching the center. This view is looking at the Canadian side where the new bridge is under the old. In the foreground, the American arch is not directly under the old bridge. The cables at the right are propped over the American arch. A number of temporary suspenders support the cables that are holding up the arch. Note the network of supporting girders and the ladder that the workers would have used to reach the working end of the arch. (John Burtniak Collection)

When the two ends of the arch approached the center, there was interference between the old and new structures due to the 14-foot (4.3 m) shift of the axis of the bridge. It became necessary to cut away 200 feet (61 m) of the suspension bridge at the center in order to close the arch. This shut off traffic for about three weeks in mid-April. Traffic during the construction period was otherwise maintained without serious interference. (NFBC Archives)

By mid-April 1898, the two arches joined at precisely the right angle, and the arch became rigid. The difficulties that were faced at this juncture in the closing of the Railway Steel Arch Bridge did not occur. This was due to the use of hinges, which allowed the arches to pivot, however slightly, to make the union of the two ends exact. The day of this union of the arches has not been recorded, but this April 17 photo opportunity for workers, engineers, and owners may have been the day. (NFBC Archives)

During the month of May 1898, the roadway was pushed out from each bank above the arch. The bow-shaped shore arms were completed in steel, as were the supports that held up the roadway. This view is of the American side. (NFBC Archives)

Early in June, it was time to wind up the cables from the old Suspension Bridge. These were pulled off their towers and wound onto spools by the use of this steam-powered engine. They would be transported north to be used in building the new Lewiston-Queenston Suspension Bridge (see Chapter Ten). The old towers were also dismantled at this time. (NFBC Archives)

This photo above from June 16, 1898, shows the roadway only half completed and without its trolley rails. The bridge did open for pedestrian traffic, however, on June 23. By June 30, the roadway was sufficiently finished to allow the first crossing by an electric trolley observation car filled with bridge and railway officials as well as some neatly dressed spectators. (NFBC Archives)

Regular electric trolley car service began the next day on July 1, 1898, even as cleanup work continued on the bridge deck. At the time, the Upper Steel Arch Bridge was the longest single-span arch in the world at 840 feet (256 m). (NFBC Archives)

The new Upper Steel Arch Bridge would go on to have an illustrious but relatively short forty-year career. Aesthetically pleasing in appearance, it became part of the fabric of Niagara Falls' tourist sights. This photograph of the recently opened bridge shows the tram lines and tram coming off the bridge. The waiting room and ticket office is to the left of the entranceway. The building with the pointed roof housed the bridge ticket office and customs inspection offices. The water falling from openings in the gorge wall is the tailrace water from the outlets of various water-powered industries that lined the gorge bank on the American side. (NFBC Archives)

These postcard shots from the early 20th century always tried to combine the Falls with the bridge. The American Falls had an unbroken crestline in those days, and the view of the bridge from Luna Island was spectacular, especially in winter. (John Burtniak Collection)

No tourist's photo album would be complete without shots of a *Maid of the Mist* vessel or the tram line that ran along what is now River Road. The International Traction Company, the owners of the Niagara Gorge Railway on the American side, purchased the Upper Steel Arch Bridge and the Niagara Falls Park and River Railway in April 1899. They already owned the rights to the new Lewiston-Queenston bridge, thus laying the foundation for the Belt Line Electric Railway system, which would carry tourists and ordinary passengers on a remarkable gorge journey. (John Burtniak Collection)

An ice jam early in 1899 caused the Bridge Company to hire engineers R. S. Buck and C. C. Schneider to inspect the abutments for damage. While there was only slight and repairable damage to the bridge structure itself, the solution to any future problem was the building of a concrete wall between and ahead of the abutments. The theory was that any ice coming down the river in future would simply lift over the wall—as it would against the prow of a ship. This concrete wall, believed to be an important added safety feature, can be seen at the base of the bridge on the American side in this July 4, 1937, photo. (NFBC Archives)

As it turned out, this concrete wall would not save the bridge in the long run. Ice bridges of various depths were a common feature in most winters of the early 20th century. On only two occasions—the first in 1899 and the second in 1908—was the bridge damaged to any extent. Both times, the damage was repaired, and new and strengthened girders near the abutments were put in place. This photo from 1906 or 1907 shows the extent of the ice bridge. The incline railway is on the left beside the American Falls. Crowds of people always ventured onto the ice for a chance to experience the thrill of being so close to the thundering water. (*Niagara in Summer and Winter*, ca. 1907)

This photo from the winter of 1937 shows the extent of the ice across the entire river. At the time this shot was taken, the ice does not look particularly safe. (NFBC Archives)

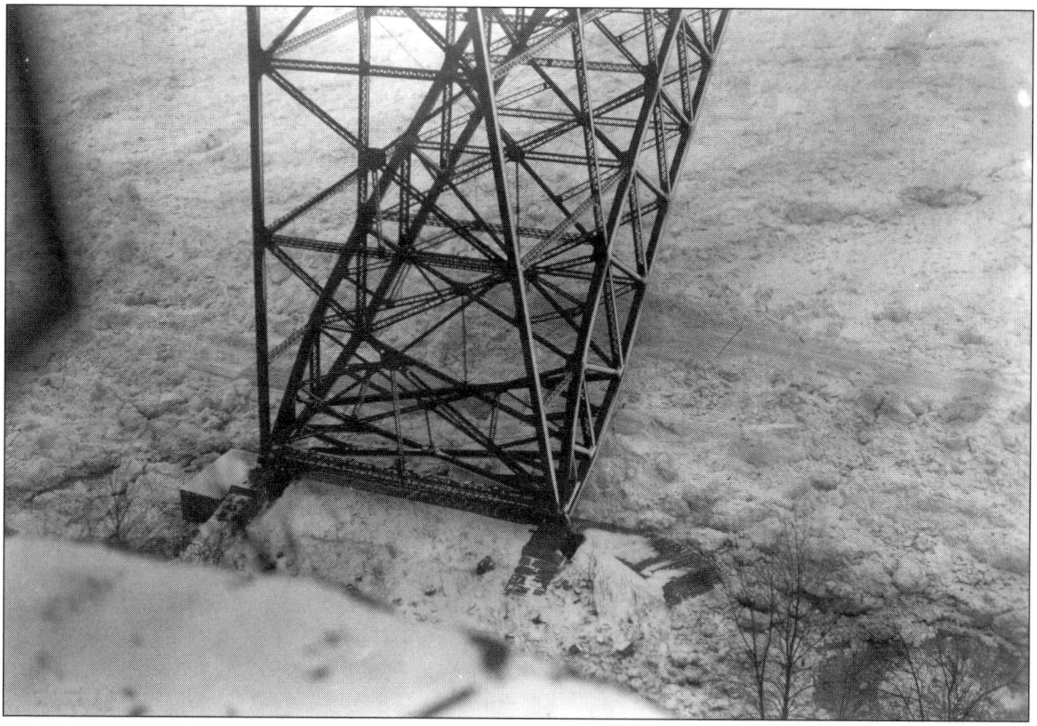

It was an ice jam of unprecedented proportions in January 1938 that caused the bridge's demise. The meteorologists had an explanation: "...thin ice on Lake Erie was followed by a January thaw, then three days of high winds. None of these elements alone could harm the bridge but all three together spelled destruction." The ice around the abutments built up quickly on January 25. A plan to dynamite the jam was abandoned. As more and more ice poured over the Falls, maintenance crews on January 26 could only watch as the ice grounded on the river bottom, backing up the water more than 40 feet (12 m) by nightfall. Rivets began popping, and some girders that supported the arch became twisted. The doomed bridge was closed to traffic at 9 a.m., power cables were cut, and the death watch began. This photo from January 26 shows the ice around the abutments on the American side. (NFBC Archives)

Hundreds of photographers and journalists descended to the bridge site, ready to file exciting stories. Their wait must have been interminable. Droves of local spectators were also drawn to the bluffs on either side of the entranceways to witness the impending destruction. This shot is from the American side just before the collapse. (NFBC Archives)

The end of the bridge came suddenly, just after four p.m. on January 27, 1938. The ice pushed the weakened American abutments off their moorings. The end section went first, crashing down the bank, and then the center section folded, pulling the end section from the Canadian side with it. In just a few seconds, the entire bridge lay on the ice like a beached whale. A number of photographers claimed to have caught the bridge as it fell. Photographs of the actual fall, however, are blurred and unreliable. This long view from Goat Island shows the remains on the ice. (John Burtniak Collection)

This close-up postcard view shows the extent of the ice in the river at the time of the bridge's collapse. This particular ice jam of 1938 would also cause considerable damage northward as far as Queenston, Lewiston, and Niagara-on-the-Lake. The downed bridge lying helplessly on the ice became the object of special tourist excursions to the area in the weeks that followed. Photos appeared in the worldwide press, and hundreds of postcards made their appearance as dealers capitalized on the event. (John Burtniak Collection)

Salvage operations began almost immediately from the Canadian side. On February 5 and 6, the bridge was dynamited into sections. Attempts were made to remove the tons of wire strung beneath the flooring, the property of the Niagara-Hudson Power Company and the Bell Telephone Company. Steel began to be lifted from the gorge by the Robertson Construction Company of Niagara Falls, and by February 19, they had recovered over three hundred tons. Salvage on the American side was impossible because the ice was too weak to support machinery. This photo from February 1938 shows the salvage crane and the blocked-off former entranceway on the Canadian side. (NFBC Archives)

The pieces of bridge sat immobile on the ice through February and March 1938. On April 11 and 12, the ice began to break up, carrying the pieces downstream with it. The last piece, shown here in front of the Niagara Falls Power Company, sank beneath the water late in the afternoon of April 12—a full seventy-five days after the collapse. The large square structures above on the left are the cooling towers of the Pittsburgh Reduction Company, forerunner of the Aluminum Company of America. (NFBC Archives)

All blame for the disaster was laid upon the extraordinary ice conditions of the winter of 1938. Richard Buck, the assistant to Leffert L. Buck at the time the bridge was being designed in 1896, offered his comments in April 1938. He stated simply that ice did not enter into the calculations of the engineers at that time. Commenting further on the loss of the bridge, the long-retired Richard Buck stated: "...the old pioneer rendered a good account of itself in providing pleasure and service and in earning many times over its cost for the owners. It went down fighting against hopeless odds and without the loss of human life." One interesting artifact remains today at the bottom of the gorge on the Canadian side—a piece of the limestone abutment with the remains of the hinge bolt, with the roller bearings totally rusted but still attached. (Photo courtesy of George Bailey)

Eight

Rainbow Bridge

From the very early months of 1938, there was no doubt in the minds of businessmen, tourists, and especially legislators on both sides of the border that the Honeymoon Bridge needed replacement as speedily as possible. It was not long, however, before complications arose. International bridges came under federal control, and none were publicly owned. Previous governments were content to leave the building to private capital. The International Railway Company, a consortium of Buffalo businessmen and the owners of the collapsed Honeymoon Bridge, quickly announced their intention to replace the structure. This intention was quashed, however, by an early Roosevelt Congress that revoked the law granting charters to private companies for international bridges. The way was thus paved for provincial and state legislators to take action. The Niagara Parks Commission also quickly brought forward its already existing plan to build a new bridge. Government bills were brought forward at Ottawa and at Washington to establish a new body to build the bridge. President Roosevelt signed the bridge bill in June 1938, creating the Niagara Falls Bridge Commission—an organization with the authority to build and operate a bridge in cooperation with Ontario. In April 1939, by an Order-in-Council, the Canadian Government gave the Commission, thus Ontario, the authority it needed to proceed. What followed was more than a year of jurisdictional land problems and lengthy debate over whether a publicly owned or a privately built bridge was best suited to the situation. This all ended with a healthy settlement for the International Railway Company from the Niagara Parks Commission, and the acquisition of the necessary lands. Construction began in May 1940 and was completed on November 1, 1941. In 1942, the Rainbow Bridge, as the Bridge Commission had dubbed it, won an award from the American Institute of Steel Construction as "the most beautiful bridge." Thousands of photographs were taken of the new structure, but perhaps it was one such as the above that influenced the award-granting panel. (NFBC Archives)

In December 1939, the Commission signed a contract for the sale of over four million dollars' worth of bonds (subscribed entirely in the U.S.) to finance the project. Plans for the expropriation of properties at the bridgehead on both sides of the border went forward at the same time. Most active in the multitude of negotiations in 1938 and 1939 was Thomas B. McQuesten, vice chairman of the new Bridge Commission, Ontario Minister of Highways, and chairman of the Niagara Parks Commission. It was he who was primarily responsible for the selection of the design for the bridge and its approaches.

In 1920, a Canadian-born engineer and bridge builder, J. A. L. Waddell, presented a paper to the American Society of Civil Engineers where he unveiled a sketch that closely resembled the eventual design. Waddell had established an engineering firm in New York and had hired a young Shortridge Hardesty. Subsequently, Waddell's firm (Waddell & Hardesty) built a number of bridges, including the one to Grand Island that opened in 1935. There, McQuesten met Hardesty and was totally impressed. In 1937, Hardesty was retained by the Niagara Parks Commission to design the new bridge. He would later be joined by consulting engineer Edward Lupfer and architect Aymar Embury, who turned Hardesty's plans into reality. It was Lupfer's Buffalo company that designed the Peace Bridge between Buffalo and Fort Erie, which had opened in 1927.

Following test borings early in 1939, a decision was made by the Bridge Commission to site the new bridge about 100 feet (30 m) north of the abutments for the former bridge. The site itself was dedicated by King George VI and Queen Elizabeth in a sod-turning on their Royal Visit to Niagara on June 7, 1939. The official opening of the Queen Elizabeth Highway was the main reason for the Royals' trip, but the opportunity to cement the Commission's case for a new bridge at the time of serious negotiations with the International Railway Company was too good to pass up. The stone monument, shown above, was originally in the new plaza wall to commemorate the event. It was later damaged in an automobile accident, then stored, repaired, and moved in 1999. It now rests under the bridge along the riverside pedestrian walk. (NFBC Archives)

With Canada already at war in 1939, but the United States still neutral, there was concern about the future availability of quality steel. For this reason as well as for aesthetics, concrete was chosen for the approaches and supporting pillars. In March 1940, contracts for the substructure were awarded to the McLain Construction Company of Buffalo and Aiken and MacLachlan of St. Catharines. The superstructure contract was awarded to Bethlehem Steel of Pennsylvania. A subcontract for the steel bracing went to the Canadian Bridge Company of Windsor, Ontario. The expropriation and clearing of properties at the bridgeheads proceeded through the spring. Ground was finally broken on the Canadian side on May 4, and in a more formal ceremony on the U.S. side on May 16, complete with an Army band. In this photo, T. B. McQuesten, Ontario Minister of Highways and now chairman of the Bridge Commission, poses with Samuel M. Johnson, vice chairman. They used the two-handled shovel, a symbol of friendship and cooperation, to turn the first sod. (NFBC Archives)

The photo below, from early 1940, shows the excavations deep in the rockface of the gorge for the abutments on the American side. These were dug into the Clinton limestone rock layer. (NFBC Archives)

Because of the depth and turbulence of the river, supporting falsework was not feasible. The answer lay in the construction of six massive concrete support columns on each side of the river. These resembled the wooden falsework of earlier bridges, but they were permanent, and the three arches that resulted became a significant part of the general aesthetics of the new structure. In the photo above, four columns had been completed, and the footings were in place for the remaining pair. The tallest of these would rise 120 feet (36.6 m). Below, the columns are complete and the three arches are taking shape. All the concrete for the project was mixed on site. These photos are from September and October 1940. (NFBC Archives)

At the base of the columns closest to the river, huge skewbacks or abutments were constructed. Onto these skewbacks the first members of the steel arch would be bolted. Each one contained thirty-two anchor bolts, seen here set in the concrete inside their temporary wooden framing. (NFBC Archives)

In this early December 1940 photo, the area between and above the arches is being closed in. It would form a base for the heavy cranes about to arrive. Notice the ladder on the front column, the man standing beside it, and, bottom right, the bolts set into the skewbacks. (NFBC Archives)

During the winter of 1940–41, work slowed somewhat due to cold weather. This view of the U.S. side shows the completed arched approach tower awaiting the first steel. There was very little ice on the river that winter, and nothing compared with the winter of 1937–38. (NFBC Archives)

The first steel arrived early in February 1941. The installation of the first grillage was a photo opportunity for the commissioners. The grillages were rectangular sections of fabricated steel weighing sixty tons each, measuring 22 by 12 feet (6.7 by 3.6 m), and bored precisely with holes to fit over the anchor bolts. The 32-foot (9.8 m) anchor bolts were long enough to pass through the grillages and into the first rib sections of the arch. The commissioners all climbed down the 494 steps to the site where Chairman McQuesten christened the February 17 occasion with a bottle of Niagara River water. (NFBC Archives)

To begin the construction of the main steel arch, a steel frame structure known as a "tieback bent" was built on top of the bridge's approach arches. Cables from each section of the steel arch are "bent" over the tieback and anchored to a 600-ton block of concrete buried 30 feet (9 m) in the rock behind the structure. The photo above shows the tieback bents in place, as well as the ramp that would transport the steel sections from the supply area. The photo below shows the tieback and the eighty-ton derrick that would lift the steel arch sections 150 feet (46 m) down to the work site. The first section was bolted into place on February 27. (NFBC Archives)

The main arch span of the Rainbow Bridge was engineered as hingeless. This means that the measurements had to be precise enough so that when the two sides came together, they would meet as closely as engineering skill would allow, because the main arches could not move to close the gap. The photo above shows a steel section en route to the work site, and it would later be lifted down to the arch. Below, the section is carried by traveller derrick out to the work site. (NFBC Archives)

The main arch spans consist of two steel box girder ribs spaced 56 feet (17 m) apart. Each arch is made up of twenty-four girders weighing up to seventy-five tons each. The arch sections were delivered to the head of each abutment and then lowered by an eighty-five-ton derrick into the gorge. There, they were received by a forty-ton traveller derrick that moved along the arch and set each section into place. This photo shows an arch section being moved out to the end by the traveller derrick. Steel bracing connects the two ribs. Note the safety net in place. (NFBC Archives)

As the ribs were extended, their weight was supported by cables bent over the tiebacks. This photo from April 1941 shows the two halves of the ribs approaching each other. (NFBC Archives)

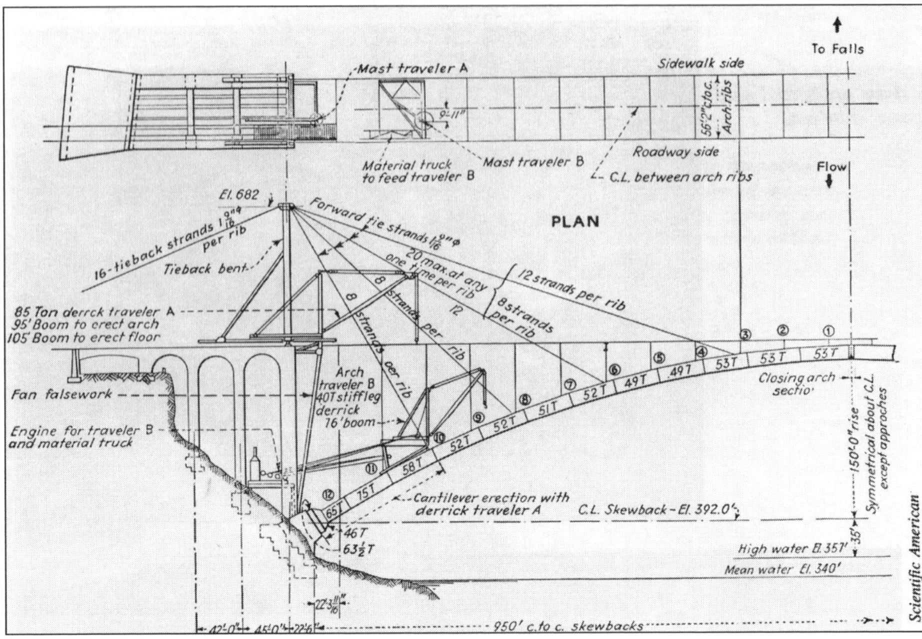

This engineering sketch shows the tieback tower, which supports the cables holding up the otherwise unsupported arch as it was built out from the foundations. A hoist under the concrete arches raised and lowered the cables. The two travellers are shown. The traveller (A) above remained stationary until the construction of the bridge deck. It lowered the arch sections to the traveller below (B), which ferried them out to the ends. (*Scientific American*, October 1941)

By early May 1941, the two sides of the arch had almost come together. (NFBC Archives)

This photo from early May 1941 shows the traveller derrick nearing the end of its work. Also shown are the mighty cables supporting the enormous weight of the arch. The last steel section, marked "Bethlehem" (bottom left), is on its way up the arm of the arch. (NFBC Archives)

This photo from a distance was taken in May. It shows both traveller derricks at the center meeting point. The two 475-foot (145 m) arch arms are within inches of each other. (NFBC Archives)

Whistles sounded and flags were unfurled on May 19, 1941, when the two ends of the bridge were completed and the last fifty-four-ton steel section was riveted into place. The two ends were a mere 11 inches (28 cm) apart and perfectly aligned. A temporary linking joint was installed. The exact dimensions of the remaining gap were then forwarded to Bethlehem Steel for the fabrication of the keystone, the last critical piece of steel. It would, in fact, support the entire span. The 11-inch gap between the two arch ends is visible below the workers. This photo from May 19 shows the temporary keystone section being lowered into position. Once the permanent keystone was installed, the arch became self-supporting upon its abutments, allowing the removal of the temporary cable supports and towers. (NFBC Archives)

The unsung heroes of the Rainbow Bridge construction were certainly the "high steel" workers. More than half of the bridge crew (of about sixty members in total) were Kahnawake Indians. Their fearlessness and unique ability made them invaluable in this precarious work. This particular ironworker gazes out at the Falls that were so well known to his ancestors. Native people like him over the years have been prominent in the "high steel" industry, particularly in New York City. It is also interesting to note that for the first time in Niagara's bridge-building history, no lives were lost to accidents during construction, although five workers did fall into the safety netting. (NFBC Archives)

With the arch completed in May 1941, work on the deck began immediately. Steel spandrel columns, resting on the arch ribs, carried the steel floor girders and concrete deck of the roadway above. The deck beams were carried out by the traveller derrick, using its 95-foot (29 m) boom to give it maximum reach. (NFBC Archives)

The deck was laced with rebar, over which the cement was poured. Concrete pouring during the summer months began in the middle of the bridge and worked toward the end on both sides. The surface was left deliberately rough to provide traction for car tires. The deck is 1440 feet (439 m) long, with two 22-foot (6.7 m) roadways separated by a 4-foot (1.2 m) center median that housed the lamp standards. On the south side of the bridge deck, a 10-foot (3 m) sidewalk facing the Falls offered a unique view for pedestrians. (NFBC Archives)

The walkway itself was christened on October 18 by a Philadelphia couple in their late eighties. It was discovered that they had spent their honeymoon in 1876 at Niagara Falls. In 1941, they were guests of the City of Niagara Falls, New York. (NFBC Archives)

The official opening of the Rainbow Bridge took place on Saturday, November 1, 1941, in the rain and mist. Just before noon, a series of aerial fireworks was exploded over the bridge. One series contained American and Canadian flags, which dropped by parachute. Next, cars with all the Niagara Falls Bridge Commissioners had the honour of being the first across the bridge. Starting from the American side, they proceeded to the center of the bridge where, in unison, Chairman T. B. McQuesten raised the Union Jack, and Vice Chairman Samuel M. Johnson raised the Stars and Stripes. The party then retreated to the Canadian side, and at high noon, another blast of fireworks signalled that the Rainbow Bridge was open for traffic. The ceremonies were kept simple in anticipation of a more extravagant gala opening once the bridge approaches were completed. Because of the Second World War, that event never materialized. (NFBC Archives)

While the bridge itself was open with the necessary customs platforms, the plaza areas on both sides of the border continued to develop. On the Canadian side, construction continued into 1943 when the 165-foot (50 m) carillon, Bridge Commission offices, and retail areas were finally completed. The bridge linked the American and Canadian public park systems. Bridge Commission chairman Tom McQuesten's dream was to use the park system as a base to create a truly beautiful entranceway into Canada. He combined his positions as Ontario Minister of Highways, chairman of the Bridge Commission, and chairman of the Niagara Parks Commission to do just that. These, of course, were separate institutions, but he often used the facilities of one to aid another. This photo from July 1943 shows one single automobile on the bridge, an indication of wartime gasoline restrictions. The Carillon Tower had just been completed. (NFBC Archives)

The whole plaza area on the Canadian side could be seen as an extension of the nearby Oakes Garden Theatre and the Oakes Gardens, which had opened in 1937. This summer 1942 view of some of the garden area beside the plaza shows the plantings well established. (NFBC Archives)

This summer 1943 photo shows the extent of the Oakes Gardens, a landscape gardening feat itself. The carillon at the plaza was finished except for its bells. (NFBC Archives)

The Canadian Customs area in the summer of 1943 was still without much traffic. (NFBC Archives)

Plans for the Rainbow Plaza were completed by October 1941. These plans included Bridge Commission Chairman T. B. McQuesten's idea of the 165-foot (50 m) carillon to complete his dream of a truly beautiful gateway into Canada. In December, the English bell foundry firm of John Taylor & Sons was commissioned to cast the forty-three tons of brass for the fifty-five-bell set. War restrictions suspended non-essential work, however, and it was not until April 1947 that the bells were finally shipped. Two gondola rail cars brought them from Halifax to the base of the tower. (NFBC Archives)

McQuesten's plan for the carillon also included a dedication inscription for the largest of the bells to read: "To God's glory and in grateful memory of our nation's leaders, Winston Spencer Churchill and Franklin Delano Roosevelt." Prime Minister Mackenzie King was deeply offended by the omission of his name, and an acrimonious dispute ensued. Below is a view of the great ten-ton Bourdon or Churchill bell resting on blocks in front of the Carillon Tower. (NFBC Archives)

The inscription on the big bell could not be changed, however. The bells had been duly installed in April before the controversy erupted. It took a year to complete the delicate cabling of the bells to their clappers, the delicate balancing with the keyboard, and the testing. The bells' first presentation of music took place in June 1948. Here, the fifty-five-bell set is arranged at the base of the carillon, which was boarded to prevent damage to the stonework. (NFBC Archives)

The tower was closed between 1998 and 2000 during extensive interior and exterior renovations. It was fully automated in 2002, and today it provides daily chimes. This photo shows the great bell on its way up the tower for installation, lifted by a steam-powered hoist. (NFBC Archives)

This aerial photo was taken in July 1953, when traffic into Canada had become significantly heavier. The curved road, top right, was then the beginning of the Queen Elizabeth Highway. At the middle bottom, the concrete footings of the old Falls View Bridge may be seen. (NFBC Archives)

Since 1941, the Rainbow Bridge has performed exactly as the engineers said it would. Changes on the bridge and its approaches, however, have been a regular occurrence over the years, along with the regular painting and maintenance that every bridge requires. By the 1980s, ever-increasing traffic was challenging the bridge's capacity and its systems. This led the Bridge Commission to launch an ambitious revitalization program in the 1990s, which would produce a complete remodelling and reconstruction of both Canadian and American border facilities. On the U.S. side, construction of a 600-foot (183 m) glass-and-steel arc building arose above the new toll and inspection stations. It was completed in thirty months and opened in October 1998. It housed the U.S. Customs and Immigration Processing Center. On the Canadian side, the plaza was also completely transformed. Renovations included new inspection booths, new administration offices, new towers accessing the curved overhead glazed viewing gallery, and a reworking of the River Road walkways. Dedication ceremonies were held in October 2000. The photo above shows the new U.S. Customs area, while the photo below shows the Canadian plaza area as it existed in 2008. The curved, glassed 500-foot-long (152 m) Promenade above the tollbooths was opened officially in May 2001. (NFBC Archives)

Nine

Lewiston-Queenston Suspension Bridge
(Queenston Suspension Bridge)

The builder of this second suspension bridge to cross the Niagara Gorge was Edward W. Serrell. He was one of the four men in iron-age America in the 1840s who believed in the feasibility of suspension bridges, even to carry rail traffic. The others were John Roebling, Charles Ellet, and Samuel Keefer, who all built Niagara Gorge bridges. Serrell's Niagara bridge, only a pedestrian and carriage structure, was completed in 1851, three years after Ellet's bridge (see Chapter One) and four years before Roebling's railroad bridge carried its first locomotive (see Chapter Two). Serrell's bridge connected the villages of Queenston and Lewiston, and at the time it claimed to be the world's longest at 1043 feet (318 m). With Roebling's assistance, it was stabilized with guy wires in 1855 after a gale damaged the structure. These guy wires served their purpose until the winter of 1864 when, to prevent them from being destroyed by ice floes coming down the river, they were detached and lifted out of harm's way. Unfortunately, when the danger had passed, no one remembered to refasten them. Another winter gale on February 1, 1864, destroyed the structure except for its towers and main cables. The fragments hung in the breeze for the next thirty-five years before they were replaced by Leffert Buck's suspension bridge at the same site in 1899 (see Chapter Ten). This earliest known lithograph, above, shows the bridge in its lower Niagara Gorge setting below the escarpment. The ferryboat in the middle of the river continued in business after the bridge was opened. (Royal Ontario Museum)

Edward Wellman Serrell (1826–1906) Bridge Builder

Edward Serrell was born in England and came to America when he was four years old. His first schooling was in New York City, where his youthful interest gravitated to engineering. He learned some basic principles from his father and older brother, and for a time, he attended the U.S. Military Academy, the only school in the United States before 1850 to offer engineering training. From 1845 to the time of the Civil War, Serrell was engaged in a variety of railroad and bridge designs. In 1845, he was an assistant engineer with the Erie Railroad, and then he became assistant to the topographical engineers in the U.S. Army. In 1848, he was involved in the survey expedition in Panama that was locating a route for a future canal. Later that year, along with John Roebling, Charles Ellet, and Samuel Keefer, he submitted his design for the first Niagara Falls bridge. That contract, however, went to Ellet (see Chapter One). Undaunted and just short of his twenty-fourth birthday, Serrell submitted his plans for a suspension bridge downstream at Queenston. These were accepted by the companies chartered to build it, and his bridge was completed in March 1851. In 1852, he built Canada's first suspension bridge across the St. John River in New Brunswick with a cable system almost identical to the one at Niagara. In 1853, he produced another notable plan, a 3400-foot (1036 m) wire suspension bridge across the St. Lawrence River at Quebec City. This proposal was not accepted. During the American Civil War, he raised a successful engineering regiment for the Union forces. After the war, he prospered as a consulting engineer, offering advice on railway, canal, and bridge building projects throughout the United States. Serrell's designs in the late 1840s and 1850s were creative enough to place him among the leaders of American engineering innovation.

The need for a bridge in the Lewiston-Queenston area had long been felt by the local population, and there had been aborted bridge projects in the 1820s and 1830s. In the 1840s, it is likely that William Hamilton Merritt considered the Queenston site before moving upriver to head what became the first Niagara crossing—Ellet's bridge. It was not until 1849 that the Queenston Suspension Bridge Company received its charter to raise private capital to construct a bridge. With Edward Serrell as their chief engineer, construction began early in 1850, starting with clearing the site for the masonry towers. The towers, 8 feet (2.4 m) square, were only 14 feet (4.3 m) high. They were built high up on the cliffs because there was no space for them below, as this close-up from the lithograph shows. Brock's monument is visible on the top of the escarpment, upper right. The distance between the towers was 1043 feet (318 m), but the roadway suspended from the towers was only 849 feet (259 m) abutment to abutment. (Royal Ontario Museum)

The cables came from the Roebling wire rope factory in New Jersey and were assembled on the American side. No. 9 gauge wire was chosen, 150 strands to a cable. A set of five of these constituted each of the two main cables. The first was taken across in April by ferryboat. Subsequent cables were hauled across by a horse-operated capstan. They were then dragged over the saddles on the towers and anchored deep in the rock behind. Supporting the bridge deck were iron-rod suspenders attached to the main cables. The roadway was 20 feet (6 m) wide, which provided space for two carriages to pass each other, and a raised footpath kept pedestrians out of the way. Guy wires anchored to the shore walls kept the bridge steady. (John Burtniak Collection)

In January 1855, a winter gale inflicted some damage. Fortunately, John Roebling was in the area completing his own bridge upstream. He came to the rescue by supplying a system of guy wires under the floor of the bridge and anchoring them to the shores on both sides. These "stays," as they were called, served to hold the span against the uplift caused by strong winds. This early 1880s view from Queenston Heights looks north toward Lake Ontario. A steam ferry is leaving the Lewiston dock, and a sailing craft sits at the Queenston dock. On the far right of the engraving is the remains of Serrell's suspension bridge, which was destroyed in 1864. (*Picturesque Canada*, 1885)

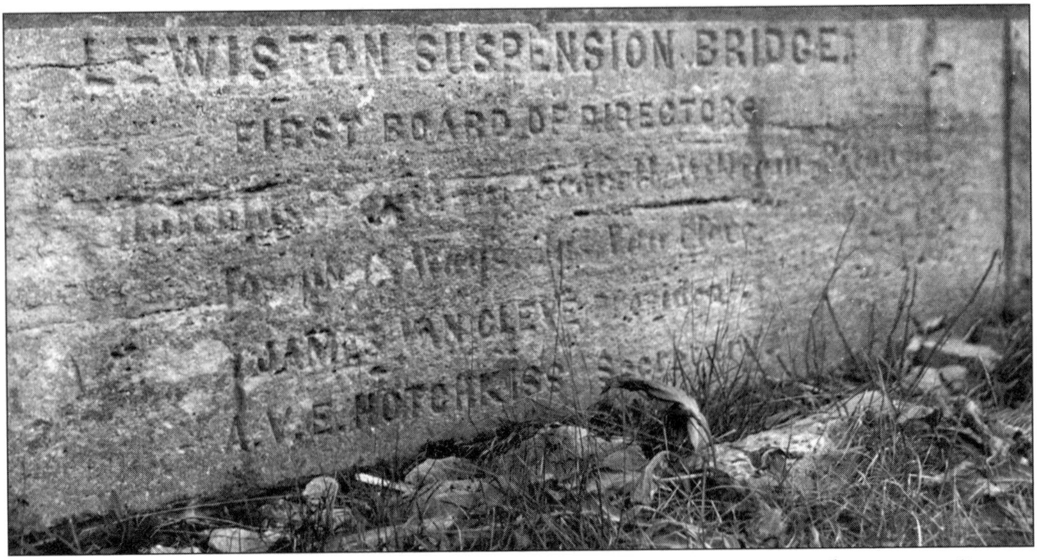

Serrell's bridge opened for business on schedule in March 1851. On March 19, according to the *Niagara Mail,* 100 Sons of Temperance marched across the bridge, along with many carriages and hundreds of pedestrians. Once across, there were three cheers for Serrell, his wife and infant son. Dinner, drinks, and speeches after the crossing were held at a Queenston hotel, and a similar party was held at Lewiston the following week. The unfortunate destruction of the structure in 1864 accounts for the shortage of images available; however, a stone plaque from the bridge was photographed in the 1950s as the lettering was fading away. It lists the first Board of Directors of the original Lewiston Suspension Bridge Company, including James Van Cleve, President. (NF NY PL)

This 1895 photo shows the ruins of Serrell's suspension bridge still hanging in place after its destruction in 1864. Some deck fragments are still attached to the iron-rod suspenders. The original set of five cables that constituted each of the two main cables is clearly visible. These would prove useful in the construction of the new Lewiston-Queenston Bridge during 1898, after which all the remnants of the old bridge were removed. (NF NY PL)

Ten

Lewiston-Queenston Suspension Bridge
(Belt Line Suspension Bridge)

The second Lewiston-Queenston Suspension Bridge fared considerably better than its predecessor. It lasted sixty-three years without mishap, and it was replaced in 1962 with the new Lewiston-Queenston Steel Arch Bridge. The contract for its construction was awarded to the Pencoyd Iron Works and Bridge Company early in 1897. Pencoyd had attained an international reputation for the quality of its iron and steel, and it had become a leading American bridge producer in the 1890s. This company had already won the contract for the Honeymoon Bridge, which would be completed in June 1898. Pencoyd's presence at Lewiston-Queenston was a natural fit. Not only were they building a Leffert Buck–designed showpiece adjacent to the Falls, they also possessed the expertise and machinery to dismantle Keefer's old Suspension Bridge as soon as the Honeymoon Bridge opened (see Chapter Six). In some ways, the new suspension bridge was an adjunct to the Honeymoon Bridge. Construction photographs abound for the Honeymoon Bridge but are non-existent from the Lewiston-Queenston site. This suggests that while they took the new bridge project seriously, the company was far more enamoured of the complexities of its Honeymoon project. Almost eighty percent of the materials used in the new Lewiston-Queenston bridge were recycled from the old suspension bridge. As construction got underway at the Honeymoon site, the International Traction Company again chose Leffert L. Buck to design the new structure. Richard S. Buck was chosen to be the constructing engineer and to supervise the dismantling of the old Falls View suspension bridge. Work on the tower foundations was completed in the spring of 1898. The new bridge would open in July 1899. The photo above was taken soon after. (NF NY PL)

Richard Sutton Buck (1864–1951) Bridge Builder

Kentucky-born Richard S. Buck studied civil engineering at the Rensselaer Polytechnic Institute, graduating in 1888. After a brief stint in the U.S. Army, he worked with Leffert L. Buck's firm of consulting engineers in New York. Although he was not related, he was heavily influenced by his employer and mentor. As resident engineer on Leffert Buck's Niagara bridge projects in the 1890s, he gained valuable experience in the construction of both the Lower Steel Arch Bridge (opened in 1897) and the Upper Steel Arch Bridge (opened in 1898). Niagara was his bailiwick, in fact, in the l890s. With Leffert Buck's blessing, Richard was chosen by the International Traction Company of Buffalo to supervise the building of the new Lewiston-Queenston Suspension Bridge, using recycled materials from the old suspension bridge, which was dismantled carefully in the summer of 1898. The new bridge opened as planned and on budget in July 1899. From 1901, Buck served as chief engineer in charge of the design and construction of the Manhattan Suspension Bridge crossing the East River in New York City, which opened in 1909. He then became the chief engineer for the Dominion Bridge Company of Montreal. During the First World War, he commanded a U.S. Army engineering battalion. Following the war, Buck established his own consulting firm in New York and finished his career with the Public Works Administration in Washington, D.C.

Survey work was undertaken in 1896 for the new bridge, which would be built on the same site as Edward Serrell's suspension bridge (destroyed in 1864). The rebuilding of the towers followed in 1897 and into 1898. These were larger and more substantial than the towers from the 1850–51 bridge, since they were required to support heavier cables. On the U.S. side they were 26 feet (7.9 m) square and 28 feet (8.5 m) high. On the Canadian side they were set higher up on the cliff so were slightly smaller and not as high. This postcard view is looking toward the American side. It shows the huge masonry wall that had to be built to buttress the shale layer upon which the tower was built. (John Burtniak Collection)

This is a rare close-up view of the stonework for the foundation on the Canadian side that would support the stone tower. It was taken in February 1898. Across the river, the level terrace for the Gorge Railway line is visible. (John Burtniak Collection)

Cables for the new bridge were strung over the saddle fixed to the top of each tower. They were then attached to the eyebars below and set into anchorages behind the towers. The anchorages were drilled deep into solid rock and heavily cemented. (Ontario Archives)

The old cables from the Falls View Suspension Bridge were wound onto their reels at the Honeymoon site in June 1898, just as the new Upper Steel Arch Bridge neared completion. These cables were cut in half and doubled to provide twin main cables on either side of Richard Buck's suspension bridge. When measured at the Lewiston-Queenston site, they were too short, so a series of eyebars was added to make up the shortfall at each end. This view, looking to the American side, shows the connecting eyebars attached to the upstream pair of cables. (John Burtniak Collection)

As reported in the *St. Catharines Standard*, the bridge opening on July 21, 1899, was accompanied "with a profusion of international bunting, champagne, artillery and an unlimited supply of Anglo-Saxon cordiality"—all accompanied by a U.S. regimental band. A trolley rolled out from each end, carrying the dignitaries. They all met under an arch at the center of the bridge for an exchange of congratulations, toasts, and speeches before heading to Queenston Heights Park for a sumptuous lunch under Brock's Monument. The arch is visible in this photo at the spot marking the international boundary. (NF PL)

The new bridge stretched 1040 feet (317 m) tower to tower. The roadway deck itself was 800 feet (244 m) long and 25 feet (7.6 m) wide, and it was secured by guy wires attached to the embankments. Steel wire suspenders between the main cables and the bridge deck ensured stability. A single rail track ran down the center to accommodate the Niagara Falls Park and River Railway trolley line. (NF PL)

Aside from the convenience afforded to the towns of Lewiston and Queenston, the greatest impetus for erecting the new suspension bridge came from the International Traction Company, which was eager to complete the link to make a dual crossing of the Niagara River possible. This would provide a remarkable circular tour of the gorge, and had considerable potential for tourist traffic. This Great Gorge Route trolley line did operate successfully until the 1930s when the Depression and the automobile hastened its demise. (*Niagara in Summer and Winter*, ca. 1907)

This view from Queenston Heights is looking north toward Lake Ontario. It shows the edge of the escarpment near the place where Niagara Falls first plunged over it more than 12,000 years ago. The left side of the photo would be Queenston Heights. (John Burtniak Collection)

This rare 1915 aerial view, looking south, shows the lower Niagara River from Queenston to Niagara Falls. The Great Gorge Route is clearly visible as it crosses the new Suspension Bridge. Brock's Monument is also visible above the village of Queenston. (NF PL)

This is one of the last aerial photos of the second Suspension Bridge. It was taken in 1962 as part of a campaign to find a buyer. A new Steel Arch Bridge was opened on November 1, 1962, and the Niagara Falls Bridge Commission (the owners since 1959) set a March 1963 deadline for the removal of the now-redundant suspension bridge. The campaign suggested that buying the bridge would be something "for the man who has everything." (NFBC Archives)

There were no takers for the bridge, however, so the Bridge Commission had to pay for its removal. That contract went to the Consolidated Contracting Company of Buffalo, which began salvage operations the day after the new Steel Arch Bridge opened. They completed their work before the March 1963 deadline. The steel was sold to Canadian scrap dealers. This photo, from December 1962, shows the beginning of the salvage operation. Interestingly, 1962 marks the centenary of the deaths of both Catharine Merritt and William Hamilton Merritt, the couple who had first dreamed of a Niagara River bridge crossing. (NFBC Archives)

Eleven

Lewiston-Queenston Steel Arch Bridge

By the mid-1950s, voices on both sides of the border were calling for a new Niagara crossing that would reflect the developing continental nature of North American transportation systems. A new bridge would speed long-distance highway traffic and would be a major connection in the growing network of four-lane expressways. The most persistent voice in promoting the new bridge was Robert Moses, president of the Power Authority of the State of New York. His first proposal in 1956 was part of his "power, parks and parkway" plans for the Niagara Frontier. He also supported the Niagara Falls Bridge Commission's efforts to acquire all three gorge road bridges. Ontario Premier Leslie Frost was equally supportive. Details were worked out by late 1958. The Bridge Commission would finance the bridge construction, the Ontario highways department would look after the Canadian approaches, and federal and state funds would cover the American approaches. Contracts with Bethlehem Steel and the consulting engineering firm of Hardesty and Hanover were confirmed in October 1960. Ground was broken in November, and construction of the abutments began shortly after. The arches came together in January 1962, and traffic began to flow in November of that year. The new bridge was formally dedicated in June 1963—the world's longest fixed steel arch span and a winner of the "most beautiful long-span bridge" award from the American Institute of Steel Construction.

Surveys, test borings, the acquisition of property, and financing arrangements occupied everyone's attention through 1959 and 1960. The photo above shows (left to right) Bridge Commissioner Stewart MacInnes, Bridge Commission Chairman Walter Johnson, Secretary-Treasurer C. E. Kaumeyer, and Vice Chairman The Honourable Charles Daley surveying the proposed gorge site. The bridge would be located about three-quarters of a mile (1.2 km) upstream from R. S. Buck's 1899 Suspension Bridge and about six miles (9.7 km) below the Rainbow Bridge. (NFBC Archives)

November 2, 1960, was indeed a happy day when the Bridge Commission turned the first sod for the new Lewiston-Queenston Bridge. Commission members (left to right) Charles Stevenson, C. E. Kaumeyer, Chairman Walter Johnson, Vice Chairman Charles Daley, John Bingenheimer, and Stewart MacInnes took advantage of the much-anticipated photo opportunity. (NFBC Archives)

Like the ones at the Rainbow Bridge, two giant abutments (skewbacks) were constructed on each riverbank as bases for the legs of the arch. These sat on concrete pads dug deep into the cliff's rock layers. The four 10-inch (25 cm) pipes, seen in this photo protruding from the floor at an angle, were to accommodate long anchor bolts that would run through the concrete abutment and 30 feet (9 m) beyond into the rock. As it turned out, no movement of the skewbacks was detected during the construction of the bridge, so inserting the deep anchor bolts became unnecessary. (NFBC Archives)

Work on the giant abutments (skewbacks) was completed by the summer of 1961. These are the reinforced concrete bases to which the steel arches would be fixed. This photo shows the arches' anchor bolts extending out to receive the grillages. (NFBC Archives)

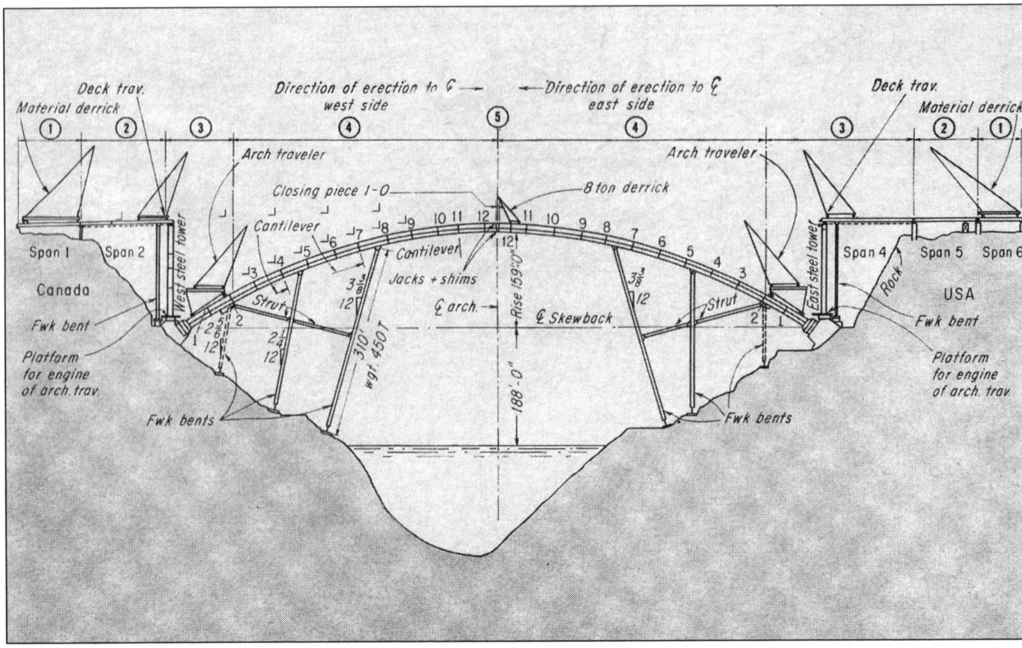

While the Lewiston-Queenston Bridge is almost a twin of the Rainbow Bridge, the method of supporting the 3000 tons of arch ribs during construction was dramatically different. With falsework impossible at the Rainbow site, 130-foot (40 m) towers were erected atop each approach pier. A cable suspension system was put in place to support the arch as it was extended (see Chapter Eight). At Lewiston-Queenston, falsework was employed and consisted of three braced two-legged supporting "bents" placed under each half of the arch. These were attached to each other by bracing struts. The diagram above shows the plan of the falsework bents supporting the arches already in place.

The diagram above also shows the general construction system. A material derrick lifts the prefabricated sections from trucks on the Canadian side, or from rail cars on the American side, and transfers them to the roadway deck. One traveller derrick moves them to the edge of the roadway deck and lowers them to the base of the arch. The second traveller derrick moves the sections out to the working ends of the arch. (NFBC Archives)

This September 1961 photo shows the derrick lowering the second section of the first fifty-five-ton spandrel column. It would be joined to the first section, which was bolted to the top of the skewback. Each section of the four precisely engineered columns had a pre-assembled working platform near the top to facilitate the connection with the next section. (NFBC Archives)

The whole spandrel column was lowered over and onto the anchor bolts, which were set into the skewbacks. (NFBC Archives)

As the spandrel column was being erected to hold up the bridge deck, the first falsework "bent" was already in place behind it. When the spandrel column was secure, the falsework bent was removed. (NFBC Archives)

In this photo, the sixty-ton steel grillages have been fitted over the anchor bolts, which protrude from the skewbacks. Each is now ready to receive its first arch rib section. (NFBC Archives)

In September 1961, the first arch rib section was lowered over its anchor bolts to attach to the skewback. In front of the arch rib is the first of the under arch falsework bents that would hold up the rib sections to follow. On the Canadian side, this falsework bent is the only one that was perfectly vertical (see diagram on page 124). (NFBC Archives)

By late October 1961, four rib sections were in place on both sides. Note the bridge workers on the wooden platform at the bottom, dwarfed by the massive steel pieces. (NFBC Archives)

By early December 1961, the arch ribs had advanced by five sections. The photo above shows the outer falsework bents under construction to support the arch as it was extended. The photo below, taken in late December, shows the arch extended to eight sections. In the distance, the old Suspension Bridge is visible near the river's surface. (NFBC Archives)

In December 1961, another seventy-five-ton rib section was lifted by the traveller derrick and lowered to its predetermined position. (NFBC Archives)

As each arch section was positioned, an air-powered winch worked through a five-line cable system to pull the rib sections together before riveting. (NFBC Archives)

The arches met on January 24, 1962. In this view, the traveller derrick is about to position the rib. The celebration flags are already in place. Note the workers inside the lower rib above the safety net. They were getting ready to receive the final rib section of this half of the arch. (NFBC Archives)

The traditional handshake (above) between an American and a Canadian high-steel worker marked the milestone. The aerial view (below), taken that cold January day, also shows the work yet to be done. Both photos show the 12-inch (30 cm) gap between the two arches, which would eventually be filled by the keystone closure section. A temporary bearing block was inserted. (NFBC Archives)

It was almost seven weeks before the highly engineered keystone piece was ready for installation. Two powerful 800-ton jacks were removed from the falsework footings and installed at the crown opening. Precise measurements of the actual thrust of both halves of the arch were checked and rechecked before the keystone was set into place, making the arch self-supporting. This photo shows the keystone about to descend to its permanent resting place on March 16, 1962. (NFBC Archives)

Rapid progress was made on the roadway deck during the spring of 1962. It was supported by spandrel columns rising from the arch ribs in ever-diminishing lengths as they approached the center. (NFBC Archives)

These photos above and below show the traveller derricks at work extending the roadway bed. When complete, the derricks would return to their respective approach spans to be dismantled by the materials handling rigs. (NFBC Archives)

This is a general view of the installation of the deck steel. The cross bracing between the arches is visible under the deck girders. (NFBC Archives)

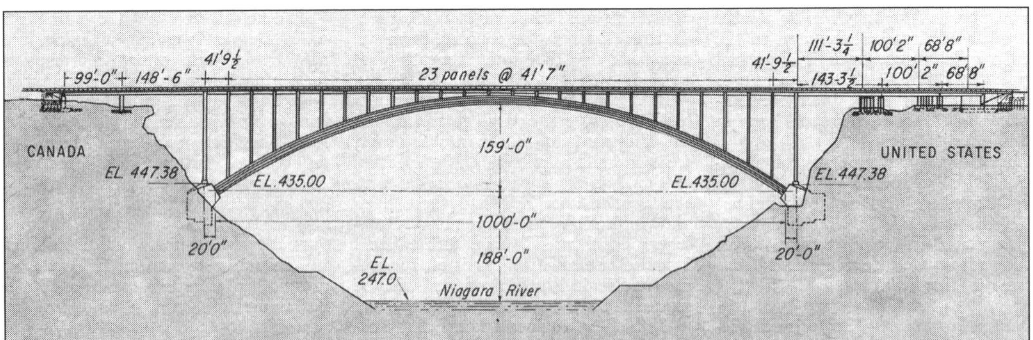

This engineering drawing shows the completed bridge project. While the Rainbow Bridge had a 950-foot (290 m) arch span, abutment to abutment, and a 1440-foot (439 m) total length, the new structure could boast of a 1000-foot (305 m) arch span and a total length, including approaches, nearing 1600 feet (488 m). The bridge had four traffic lanes and a pedestrian walkway. It now connected New York Interstate 190 with the QEW and Hwy 405, which was rushed to completion in 1963. (NFBC Archives)

The new Lewiston-Queenston Bridge opened with little fanfare on November 1, 1962. Bridge Commission Chairman The Honourable Charles Daley, beside Vice Chairman John J. Bingenheimer, cut the ceremonial ribbon at the center of the bridge. As traffic began to flow in both directions, the old Suspension Bridge was closed to traffic and was prepared for demolition. A new era in road transportation had begun. (NFBC Archives)

A formal inauguration took place at a gala event on June 28, 1963—a date set to accommodate both New York Governor Nelson A. Rockefeller and Ontario Premier John P. Robarts. There were handshakes all around, numerous speeches, and plentiful food served on the Canadian Plaza site. Here, with Nelson Rockefeller in the middle, John Robarts (left) is introduced to John Bingenheimer, the new Commission chairman. (NFBC Archives)

While the bridge itself received most of the attention, the development of the approach roadways and the plazas on both sides of the border was equally dramatic. Both plazas offered streamlined facilities designed to meet the traffic needs of the 1960s and beyond. These aerial views of the Canadian plaza (above) and the American plaza (below) were taken shortly after they opened in November 1962. (NFBC Archives)

By the early 21st century, it was clear that the expansion of traffic was straining all of the bridge's systems. Long truck queues were having an increasingly negative economic and environmental impact. An enormous and many-faceted renovation was undertaken in 2004. Called the Fifth Lane Project, $57 million was spent widening Hwy 405, strengthening and widening the bridge, improving operations and safety, and introducing a state-of-the-art traffic management system—all to ensure speedy safe passage and border security. A reversible fifth lane was dedicated for pre-approved trucks entering the United States. Renovation work began on the bridge in December 2004. After eleven months of working double shifts, six days a week, the project was completed successfully in early November 2005, ahead of schedule and on budget. These views illustrate some of the results. (NFBC Archives)

As the Fifth Lane Project got under way, plans for a new administration building were also bearing fruit. Like the bridge itself, the old administration area under the Carillon Tower at the Rainbow Bridge site had become inadequate. A new Administration Building, this time on the American side at Lewiston, was opened in September 2004 for the Operations Center and the many offices of the Niagara Falls Bridge Commission. With its cut-stone and arched entranceway, it received rave reviews from the architectural community. (NFBC Archives)

From its beginnings in 1938, the Niagara Falls Bridge Commission has been composed of eight commissioners—four appointed by the Ontario Premier and four by the Governor of New York State. These, along with the General Manager and staff, form a public authority that administers three great bridges crossing the Niagara Gorge. This photo of the commissioners was taken in April 2008. Seated, l to r: Diane Vitello; Janice Thomson, Chairperson; Norma Higgs, Vice Chairperson; Patrick Brown. Standing, l to r: Brian Wilkie; David O'Toole; Thomas Garlock, General Manager; Kenneth Loucks; Thomas Pryce. (NFBC Archives)

Twelve

Goat Island Bridges

Clustered on the U.S. side of Niagara Falls, on and around Goat Island, sit a number of small but historic bridges that have delighted tourists for almost two centuries. The State of New York acquired the Goat Island group in 1885 for a public park. The original owners, the Porter brothers, ceded it back to the state. The island was then included in the Niagara Reservation State Park, the oldest State Park in the United States. The bridges are maintained today by the New York State Park Commission.

Augustus Porter (later Judge Porter) had received the land as compensation for military service in the War of 1812. The first wooden structure was built in 1817 by the two Porter brothers. A replacement was built in 1818 and repaired in 1839 and 1849. In 1855, a cast-iron bowed arch bridge was erected (above left). It was replaced in 1900 by a 1000-foot (305 m) three-arched stone structure (above right). An additional modern steel structure, the American Rapids Bridge, was opened to the northern part of the island in 1960 to facilitate heavy vehicle traffic.

Terrapin Point on the western side of Goat Island used to boast a 45-foot (13.7 m) stone tower that offered a commanding view of the Falls. A bridge to the Point was built in 1829, and the tower was erected in 1833 by the Porter brothers who charged a fee for the use of its boardwalk access bridge to the tower. The tower (above left) was removed in 1873. On the right is the entourage surrounding the Chinese ambassador, possibly on his visit in 1868. (John Burtniak Collection)

Winter scene at Terrapin Point (John Burtniak Collection)

The old Luna Island Bridge (*Niagara in Summer & Winter,* ca. 1907)

The first bridge to the Three Sisters Islands was built in 1868. It was replaced by the present sandstone bridge in 1898. The bridges to the other two islands were replaced in 1950 and 1951 by the present concrete bridges. These stereographic views (left and right) show two of the early bridges to the Three Sisters Islands. (*Tugby's Illustrated Guide to Niagara Falls,* 1889)